Thomas Bryant

The Diseases of the Breast

Thomas Bryant

The Diseases of the Breast

ISBN/EAN: 9783743383876

Manufactured in Europe, USA, Canada, Australia, Japa

Cover: Foto ©berggeist007 / pixelio.de

Manufactured and distributed by brebook publishing software (www.brebook.com)

Thomas Bryant

The Diseases of the Breast

CLINICAL MANUALS

FOR

PRACTITIONERS AND STUDENTS
OF MEDICINE.

Fig. 1.

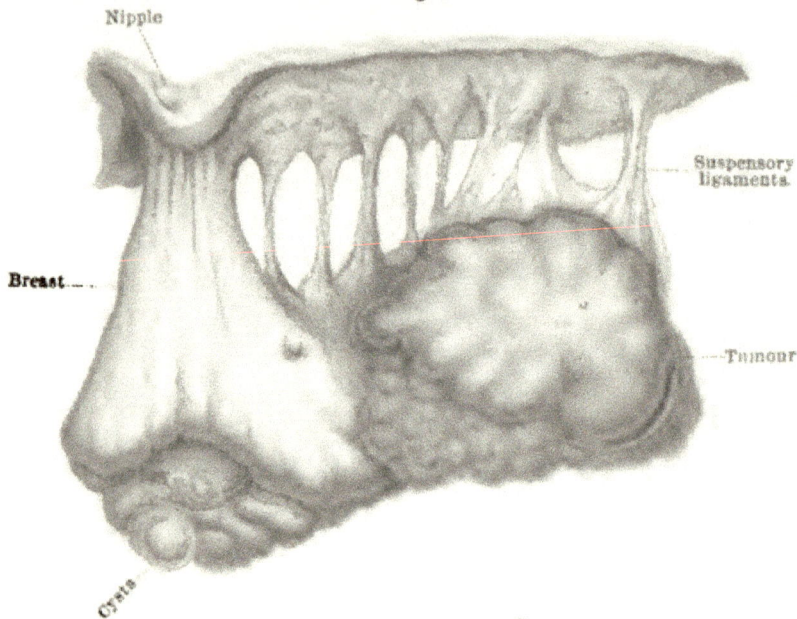

Nipple

Suspensory ligaments.

Breast

Tumour

Cysts

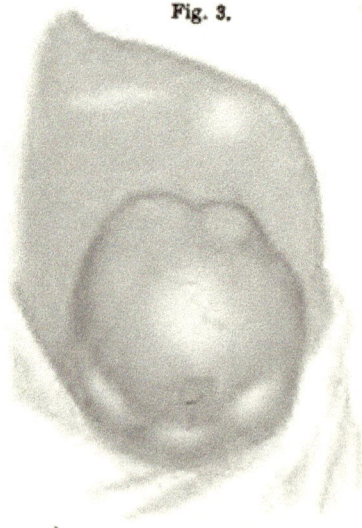

Fig. 3.

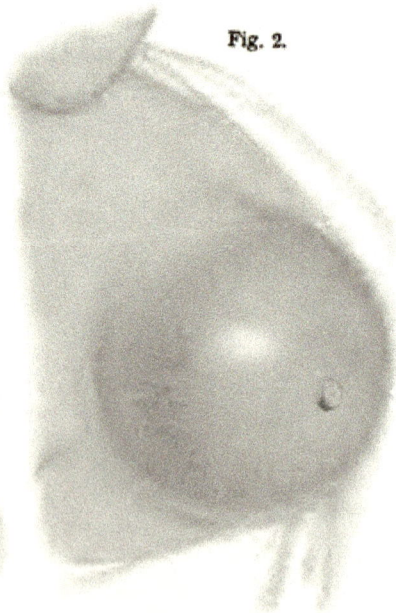

Fig. 2.

1. Encapsuled Adeno-fibroma.   3. Adeno-fibroma.
2. Submammary Abscess.

THE

DISEASES OF THE BREAST.

BY

THOMAS BRYANT,

F.R.C.S., M.-Ch. (Hon.) Roy. Univ., I.,

SENIOR SURGEON TO AND LECTURER ON SURGERY AT GUY'S HOSPITAL;
VICE-PRESIDENT, CHAIRMAN OF THE COURT OF EXAMINERS,
AND HUNTERIAN PROFESSOR OF SURGERY, ROYAL COLLEGE
OF SURGEONS, ENGLAND; CORR. MEM. SURGICAL
SOCIETY, PARIS.

WITH 13 ENGRAVINGS AND 8 CHROMO-LITHOGRAPHS.

CASSELL & COMPANY, LIMITED:
LONDON, PARIS, NEW YORK & MELBOURNE.
1887.

PREFACE.

THE object of this work is to place before the student and practitioner a clinical exposition of the abnormalities and diseases of the breast, more particularly with reference to their diagnosis and treatment. With this view the descriptions of pathological processes have not been placed in the foremost position, and it has been assumed that readers of this volume are familiar with the leading *macro*scopical as well as *micro*scopical features of the different varieties of tumours, such as are to be found described in the many excellent works on pathological anatomy.

The pathological aspects of disease have, however, been referred to, to elucidate its clinical phases, and more particularly to show how the signs and symptoms of local disease are to be explained by progressive pathological processes.

To render the clinical aspects of disease clear, brief notes of cases have been quoted, to illustrate either a subject or a symptom ; and with a like object coloured plates have been freely introduced, many of which have been copied from the originals in Guy's Hospital Museum.

A *

In preparing this work I have to thank Dr. J.
Goodhart for his kind aid in revising the pathological portion, and Mr. C. Symonds for some valued
microscopical investigations.

65, GROSVENOR STREET,
    December, 1887.

CONTENTS.

LIST OF COLOURED PLATES.

BIBLIOGRAPHY.

Brodie ; Lectures on Surg. Pathology ; 1846.
Sibley ; Med.-Chir. Trans., vol. xlii. ; 1859.
Baker ; Med.-Chir. Trans., vol. xlv. ; 1862.
Butlin ; Med.-Chir. Trans., vols. lx., lxiv.
T. Annandale ; International Encyclopædia of Surgery, by Ashhurst, vol. v. ; 1885.
H. Morris ; Med.-Chir. Trans., vol. lxiii. ; *Lancet*, vol. ii. p. 873; 1879.
A. von Winiwarter, of Vienna ; Beiträge zur Statistik der Carcinome ; 1880.
Richelot ; Des Tumeurs Kystique de la Mamelle ; Paris, 1878.
Labbé and Coyne ; Traité des Tumeurs Bénignes du Sein ; 1876.
Fovget ; Bull. gen. de Therapeute, tome xxvii. p. 359.
Conrad Langenbeck ; Nosologie und Therapie der Chir. Krankheiten, Bd. v. p. 83.
Cruveilhier ; Anat. Pathologique ; 1835-42.
C. de Morgan ; Patholog. Soc. Trans.; 1874. Discussion on Cancer.
C. Moore ; Influence of Inadequate Operations upon the Theory of Cancer, Med.-Chir. Trans., vol. l. ; 1867.
Velpeau : Traité des Maladies des Sein ; 1854. Also Translation ; Sydenham Society ; 1856.
Cornil and Ranvier ; Manuel d'Histologie Path. ; 1869.
Alfred Haviland ; *Journal of Society of Arts*, No. 1,367, vol. xxvii.; 1879.
Handyside ; *Journal of Anat. and Phys.*, vol. vii. ; 1873.
Dr. Mitchell Bruce ; *Journal of Anatomy and Physiology*, vol. xiii.
Cameron ; *Journal of Anatomy and Physiology*, vol. xiii.
C. Creighton ; *Journal of Anatomy and Physiology*, vol. xiv.
Leichtenstern ; *Virchow's Archiv*, vol. lxxiii. part 2 ; 1878.
Cohn ; *Berliner Medizinische Gesellschaft*, Feb., 1885.
Champneys ; Med.-Chir. Soc. Trans., vol. lxix. ; 1886.
Léger ; *Bull. Soc. Med. d'Amiens;* 1878.
Richet ; *Gaz. des Hôpitaux*, May 13, 1880. Traité d'Anatomie Chirurgicale ; 1857.
Ambrosoli ; *Gazetta Medica Lombarda*, No. 36 ; 1864.
Banks ; *British Med. Journ.*, Dec. 9, 1882. Harveian Society, March 3, 1887.
Dubar ; Tubercules de la Mamelle ; 1881.
Durant : *Gaillard's Med. Journ. of New York*, June, 1884.
Darwin ; Origin of Species ; 1875.
Gould ; Medical Society Proceedings ; 1884.

Sauvages; Nosologia Methodica; 1768.
Thomas; *New York Med. Journ.*, p. 337; **1882.**
Thin; *British Med. Journ.*, May, 1881.
Munro; *Glasgow Medical Journal;* 1881.
Duhring; *American Journ. of Med. Science,* July, 1883.
Wagstaffe; Pathological Society Trans., London, vol. xxvii.
Sir J. Paget; St. Bartholomew's Hospital Reports, vol. x.; 1874.
    Surgical Pathology; 3rd ed. 1870.
August Förster; Die Missbildungen des **Menschens**; **1861.**
Billroth; Surgical Pathology; 1877.
Robert; *Journ. Génér. de Med.*, tome c., p. 57.
Cæsar Hawkins; *Lond. Med. Gaz.;* 1838. Collected Works; 1874.
R. Lee; Med.-Chir. Trans., vol. xxi.
S. W. Gross; Tumours of Mammary Gland; 1880.
Sir A. Cooper; Anatomy of Breast; 1840. Illustrations of Dis-
    ease of Breast; 1829.
Manget; Dict. Ency. des Sc. Med., tome xiv.
Woodman; Trans. Obstet. Soc. Lond., vol. ix.; **1868.**
Lousier; Dissert. sur la Lactation, p. 15.
Fitzgibbon; *Dublin Quart. Journ.*, vol. xxix.; 1860.
Max Bartels; Reichert and du Bois-Reymond's Arch., 1875.
Percy; Dict. des Sc. Med., tome xxxiv. p. 525.
Sneddon; *Glasgow Med. Journ.;* 1878.
Shannon; *Dublin Quart. Journ.*, vol. v.; Feb., 1848.
Jussieu; *Lancet*, vol. xii. p. 618. Philomatic Soc.
Froreip's Notizen, bd. xxiii.
Stettegart; *Langenbeck's Archives;* 1879.
Dr. A. Henry; Statistiche Mittheilungen **über** den Brust krebs;
    Breslau, 1879.
H. Lebert; Des Maladies Cancereuses; 1851.
Birkett; Diseases of Breast; 1850. Holmes' System of Surgery;
    3rd ed. 1884.
Nunn; Cancer of the Breast; 1882.
Rudolf Maier, of Freiburg; Lehrbuch der Allgemeinen Patho-
    logischen Anatomie; Leipzig, 1871.
C. Creighton: Contributions to the Physiology **and** Pathology of
    the Breast; 1878. Reports of the Med. Officer of the Privy
    Council, 1874, 1876.
Williams, W. R.; The Influence of Sex in Disease; 1885.
Goodhart, J. F.; The Nature and Development of Cystic Tumours
    of **the** Breast; *Edinburgh Med. Journal,* May, 1872.
Bérard; Diag. différ. des Tumeurs du Sein; Paris, 1842.
Wilks and Moxon; Lect. on Path. Anat.; 1875.

DISEASES OF THE BREAST.

CHAPTER I.

ANATOMY OF THE BREAST.

IT is essential that the practitioner when examining a diseased breast should always have in his mind's eye the anatomy of the healthy organ. He should remember that the breast is a skin gland situated in the connective tissue between the layers of the superficial fascia ; that the gland itself is encapsuled, with its posterior wall slightly concave, resting and moving upon the pectoral muscle, and the anterior wall convex, separated from the skin by more or less lobulated fat, and at the same time connected with it, in suspended form, by fibrous bands, which are known as the suspensory ligaments of the breast (Plate I. Fig. 1).

The boundaries of the gland are not however always very exact, and it is not uncommon, as demonstrated by disease, to find processes, or rather lobes, of the gland projecting into parts which seem to be outside the radius of the normal gland. The frequency of this extension from its axillary border is so great, that it may be open to a question whether it may not be regarded as a normal condition. In operations upon the breast this anatomical condition is worthy of remembrance.

In thin people there is but little fat between the gland and the skin covering it, and under such

B—25

circumstances there is an absence of that **roundness
and plumpness which so** characterises the **breast of a**
healthy and well-made **woman.**

On the other hand, **in** fat women, even when **the**
gland itself from senile atrophic changes has shrunk
to very small dimensions, the deposit of fat around
the gland being large, maintains the shape and size
of the breast.

The presence or absence of fat about the gland
has consequently an important bearing in modifying
the manipular indications of a healthy as well **as** of
a diseased breast. **A small** tumour in a **thin** breast
is readily recognised, **whereas** in a fat **one it**
might be overlooked; and a scirrhous infiltration of
an atrophic breast of a fat woman might suggest the
presence of a large tumour, when in reality the bulk
of the growth is fatty tissue.

The practical bearing of the existence of the suspensory ligaments of the breast is very great, since it
is by the shortening of these ligaments that the
characteristic " dimpling" and "puckering" of the
skin, so frequently seen in the infiltrating form **of
cancer** of the breast, is brought about.

The gland itself is a conglomerate, racemose, **or**
compound organ, made up of from fifteen to twenty
distinct lobes, subdivided into lobules ; the lobules
on section presenting a pale pinkish **cream** coloured
appearance.   Each lobe is held together by firm
fibrous tissue, and yet at times is separated by fat ;
each lobule is surrounded by connective tissue for a
support, as are likewise the smaller lobules and secreting acini of the gland.

This connective tissue may become **the seat of**
connective tissue neoplasms (*sarcomata*) ; **the acini
and ducts the** seat of *adenomata* and epithelial
(*cancerous*) neoplasms.

Connected with each lobe there is an excretory

galactophorous or lactiferous duct, which terminates in
the nipple, and during lactation this tube as it approaches the nipple becomes dilated into a sinus or am-
pulla, one-quarter or one-sixth of an inch wide, and these
ampullæ act as reservoirs for the milk when secreted.
Each terminal tube opens in the nipple by a separate
orifice, which is smaller than the tube, while the small
initial tubes or ducts of the acini and lobules unite
together, and so help to form the larger tubes as they
converge towards the nipple.

The lobes, lobules, and acini of the gland are
lined by an almost structureless basement membrane
covered with epithelium, and this epithelium varies in
character according to its position and to the condition
of activity of the glands. In the lactiferous ducts,
from the reservoirs to their terminations, the epithelium appears as small oval corpuscles with a
central nucleus to each; towards the orifices of the
ducts the epithelium tends towards a squamous appearance. In the reservoirs the epithelium is of the
tesselated form; columnar cells are said also to be
present by Kölliker and Birkett. When the breast is
in an inactive condition the epithelial cells are small
and granular, but when physiological activity begins,
the alveoli enlarge, and become filled with a clear
secretion. The cells appear flattened out against the
membrana propria, or basement membrane, and contain fat globules of varying sizes. When the gland is
in full activity the epithelial cells become cubical, or
even columnar, but are irregular in size, and exhibit
indications of segmentation, and discharge into the
cavity of the alveoli.

"The secretion of the breasts is formed by the
bursting of these cells, and the discharge of their contents into the alveolus, the fat globules which were
present within the cells becoming suspended in the
fluid of the alveoli as milk globules, and the albuminous

constituents of the cells becoming dissolved and forming the casein and other proteid substances of the milk." *

"The mature functions of the breast," writes Creighton, "are in effect a sustained repetition of those cellular changes which the embryonic cells went through in order to become the breast," and no doubt the spurious or morbid pathological *evolution* and *involution* of the gland is marked by changes in its epithelial or secreting structure, which in a measure or to a degree simulate these changes.

In intermediate stages between these two extreme physiological conditions, as well as when the gland is in a morbidly active state, changes in the epithelial lining of the acini will be visible, analogous to those which take place in the gland in its physiological evolution and involution.

The blood-vessels of the gland, which are numerous, invest the alveoli, but do not come into direct contact with the secreting cells themselves. In the breast, as in other secreting glands, the vessels lie outside the basement membrane upon which the epithelium is placed, and according to Schäfer are separated from the secreting cells by the lymphatics. Hence, the material for the nutrition of the cells is not drawn from the blood direct, but from the lymph of the breast gland.

The arteries of the breast are thoracic branches of the axillary, branches from the aortic intercostals, and the intercostal branches of the internal mammary. There is a markedly free anastomosis of veins, more particularly around the areola.†

The lymphatics from the breast are also numerous; they are superficial and deep, and pass in two main lines. Those from the axillary and lower half of the gland run to the axilla; whilst from the sternal half

---

* Quain.　　　　† Haller's "Circulus Venosus."

they pass to the glands situated along the course of the internal mammary arteries beneath the sternum. Lymphatics from the upper part of the gland pass into the cervical region, and not necessarily through the axilla.

The nipple (mammilla) occupies a position slightly below the centre of the gland, on a level with the lower border of the fourth rib ; it projects in the virgin state forwards, outwards, and slightly upwards ; during the progress of a pregnancy it turns downwards. The surface of the nipple is dark, and around it is an areola, which in the virgin is of a light rosy pink colour, and in the prolific matron becomes dark ; the darkness of the areola commencing at the second month and increasing with every month of pregnancy. As the gland returns after lactation to its quiescent condition the darkness of the areola may diminish, but **does not disappear.**

The skin of the areola is wrinkled and covered with roundish elevations, which **are sebaceous follicles** known as large and small areola glands ; every gland opens by four or five ducts on each papilla. When the ducts are obstructed a sebaceous tumour may form. The nipple contains some muscular tissue, and the terminations of the fifteen or twenty excretory gland ducts. The central ducts are the larger ; and outside the ducts, in the areolar tissue which surrounds them, the blood-vessels ramify. The ducts become narrow at their orifices, and open in the little depressions seen in the apex of the nipple.

The skin covering the mammary glands receives .
its sensibility from the anterior and middle branches of the second, third, fourth, and fifth intercostal nerves, whilst the skin over the upper border of the gland is supplied by the supraclavicular branches of the cervical plexus. These points should ever be remembered, for pain may follow the course of these nerves

and be experienced over the region of the breast which they supply, without the existence of any breast disease. The mammary glands **are** the same at birth in both **sexes.** The development **of** the gland takes place **in the** female at puberty, **at or** before the first catamenial period. From puberty to the cessation of the catamenia the glands are in **the condition of** dormant developmental perfection, but this stage itself is reached only when from pregnancy **the full** physiological functions of the gland are called into activity.

After the cessation of the catamenia, when the functional activity of the glands can no longer be called into requisition, a period of functional decline commences, which in the healthy subject ends **in a** simple atrophy of the secreting acini of the gland.

At each of these periods certain troubles are prone to appear, to which **attention** will be drawn in the following chapters.

CHAPTER II.

ABNORMAL CONDITION OF THE BREAST AND NIPPLES.

THE abnormalities of the breast and nipple **show** themselves either in the way of excess **or of** deficiency. Those of excess are the more common, and strange **as** it may seem, they are more frequent in the male than **in** the female sex. Dr. Mitchell Bruce, to whom **we are** indebted for an able paper on this subject,* tells us **that** out of nearly four thousand men and women examined consecutively, sixty-one had supernumerary nipples, forty-seven were males and fourteen females. In **a** second series of cases, examined exclusively for **the** purpose, about seven out

* "Journal of Anatomy and Physiology," vol. xiii.

of every hundred consecutive individuals had the deformity, and it was found nearly twice as often in men as in women. Five out of six of Bruce's cases had but one supernumerary breast or nipple, whilst the sixth had more than one. Most of the instances observed have been situated on the front of the trunk, below the level of the ordinary mamma, and somewhat nearer the median line of the body than the normal gland. These supernumerary parts were distributed, as ingeniously pointed out by Bland Sutton, in the course of the internal mammary and deep epigastric vessels. In exceptional instances these excesses have been found in extraordinary positions. Again, these supernumerary parts are more frequently found on the left than the right side of the body, in the proportion of sixty to forty per cent. The figures of Bruce, Leichtenstern, and others, clearly indicate this fact.

Leichtenstern, in discussing the cause of this curious disproportion, maintains that the left mamma is on an average more developed than the right, and mentions Hyrtl's explanation of this inequality, that it is so because mothers use the left gland more in suckling. He, however, himself maintains that the same condition holds good in the child and virgin, which, if a fact, adds Bruce, is probably an instance of variety "as the inherited effect of use,"* and a point of extreme importance in the question of the significance or origin of supernumerary mammæ. "We may doubt also," wrote Darwin, "if additional mammæ would ever have been developed in both sexes of mankind, had not his early progenitors been provided with more than a single pair."

The supernumerary nipple is commonly placed about three inches from the normal nipple. In exceptional cases it may be only about one inch distant,

---

* Darwin, "Origin of Species," p. 8. 1875.

whilst in others it may vary from three to six inches. But whatever may be the distances in childhood, Bruce has satisfied himself that the distance may increase considerably with age.

The characters of the supernumerary gland or nipple vary exceedingly. "Some examples," writes Bruce, "might be chosen for description, which in every particular, except size, resemble an **ordinary** male mammilla, while others require considerable experience for their discovery and identification. The best-marked cases present the central papilla or nipple proper, a pigmented areola, follicles, hair, and a distinct depression on the apex of the papilla. Examples such as these are, however, very rare. The more usual condition is the presence of a small papilla, or elevation more or less like the ordinary male mammilla, "or a low ovoidal prominence, with its long diameter in the transverse direction, slightly inclined downwards and outwards, and having its **summit** usually distinctly cleft into two lobes by an opening or deep groove parallel with the diameter, this variety somewhat resembling the retracted nipple of some female breasts."

The papilla of the supernumerary mammilla may frequently be found erectile.

The areola of a supernumerary nipple may be entirely absent, or it may be represented by **the** smallest line of pigmentation around the base of the papilla. In some cases it exists as an areola of natural dimensions and appearances; in others it is pigmented in variable degrees, whilst in some examples all pigment is absent.

Follicles and hairs are not commonly present in the supernumerary nipples. In exceptional examples, however, some strong pale or black **hairs are** met with.

Supernumerary nipples found above the normal

mammilla are very rare.   Bruce says only four cases of
the abnormality are on record, and in all of these
four the supernumerary was placed outwards as well
as upwards from the normal organ.   Dr. Fitzgibbon
has recorded an example of this kind, in a soldier,
aged twenty-four, a native of Jamaica, who not
only had two supplementary nipples, one on each
breast about one inch below the normal nipple, but
two pigmentary deposits, one on either breast higher
up above the ordinary mammillæ, which were clearly
supernumerary nipples.   When the man was young
the lower supplementary were larger than the natural
nipples.   When seen by Dr. Fitzgibbon they were
smaller.

Amongst the cases of supernumerary breast which
have been recorded, there are four or five in which
five mammary glands existed; one of these has been
minutely described by M. Gorré, as reported by
Percy.   Four of the glands in M. Gorré's case were
very projecting, full of milk, and arranged in two
lines; the fifth breast was not larger than the breast
of a girl before puberty, and was placed below and in
the middle of the body, between the inferior pair, and
five inches above the umbilicus.

Cases in which four breasts have been seen are
recorded by Sir A. Cooper, Dr. Robert Lee, and
Dr. Shannon; others in which three breasts existed, by
Dreger, Bartolin, G. Hannæus, J. Borel, Robert,
and Sneddon.   Where these abnormalities exist others
are not unfrequently associated with them.   Thus,
in a girl, aged six, whom I saw in September, 1883,
with a supernumerary nipple and areola one inch
below the left nipple, there was some malformation
of the genital organs; apparently no vagina, and a
clitoris as large as a boy's penis, with prepuce and
orifice as of a urethra, which passed down the clitoris
for half an inch; the urethra was normal.

Percy states that Anne Boleyn, who had six fingers and six toes, had three breasts.

Supernumerary glands are sometimes active ; and as a rule it has been from their functional activity during pregnancy or lactation that their presence has been made known.   Dr. Bruce had not, however, seen an example of this kind, and in not a single instance would the subject of the abnormality confess to any alteration of the parts in pregnancy or during menstruation.

On April 6, 1831, Mr. Roberts, an able practitioner, exhibited before a London Society (the Hunterian) the model of a female breast with two nipples.   They were a quarter of an inch apart, and the woman was able to suckle by either.   It is said, indeed, that in most of the examples recorded when the breast is lateral, the supernumerary nipples emitted milk, but when it is median, this is not always the case.   In an instance which has passed under my observation, this was not, however, the condition ; the case was as follows : *Supernumerary nipple and breast.*   Sarah P., aged forty, the mother of eight children, all of which she had suckled, came under my care on September 25, 1862, with a supernumerary nipple about two inches below her right breast.   The nipple looked quite like the ordinary nipple, but never discharged milk.   It did not appear to have any connection with the right breast, but the parts beneath it were full, and this suggested the presence of a supernumerary gland.   The fact, however, that during her eight pregnancies no milk was secreted proves the contrary.

This deformity, like others, is at times hereditary, but not being a self-evident one, the fact of its being so is not often to be made out.   Both Bruce and Leichtenstern failed to find evidence in any of their cases of heredity, and it is mentioned in only

seven out of the ninety-two cases recorded before
these authors investigated the subject afresh. M.
Robert has, however, reported the case of a **woman**
with a supernumerary breast, whose mother had the
same anomaly, and Dr. B. Woodman saw a mother
and daughter each of whom had three nipples. Dr.
Handyside has reported a **case where two brothers
had four** breasts each, the two supernumerary **glands**
being placed below the normal mamma, but nearer
the median line of the body.

Supernumerary mammæ are found at times in
strange places. Two instances of supernumerary
glands have been recorded by Leichtenstern as having
existed upon the back. Both have, however, been
taken from old works ; one from Paulinus, and the
second quoted by Helbig. Leichtenstern **accepts the**
cases as true, and upon his authority I **name them.**
Klob has recorded one case in which the mamma was
placed over the acromion process of a man, and was
the size of a walnut ; it had also a good nipple. In
M. Robert's case the abnormal gland was placed " on
the outer part of the left thigh, four inches below the
great trochanter, **and** was about the size of a lemon.
It was the seat of pain, and of sensations like those
of the normal breasts, at the catamenial periods. The
possessor of it had suckled several children with the
third breast. The mother of the woman, aged fifty,
had three breasts, all in the chest." M. Jussieu **reports**
the case of a woman who had a third breast in the
groin, **with** which lactation was performed. " I know
of no case in literature," adds Leichtenstern, " where
accessory breasts or mammæ have been met with below the inferior border of the ribs upon the abdomen."
Bruce has, however, recorded in his paper two, if not
four cases in which it **was clear** that the supernumerary nipples had an abdominal position.

The existence of axillary mammæ cannot be

doubted. Leichtenstern has recorded five, one of which he saw. It was in the left axilla, and had associated with it a supernumerary nipple below and internal to the normal one of the same side.

A. Förster has recorded a case in which the supernumerary breast was in the axillary region and the seat of carcinoma.

A. H. Cameron recorded* a case of a pregnant woman who had an axillary swelling, in which there was a small orifice, but no nipple from which milk could be squeezed. Cohn also, at a meeting of the Berliner Medizinische Gesellschaft in Feb., 1885, exhibited a patient precisely similar to Cameron's. Cameron, in his paper, suggests that if morphological theories fail to explain the abnormality which he had described, we may accept Laycock's view that the mamma is simply an enlarged and highly developed sebaceous gland, and might make its appearance in any part of the body.

Since the above remarks were written, Dr. Champneys has described, in an able paper on the development of mammary functions of the skin of lying-in women,† thirty cases of what he has termed milk-producing tracts of integument, or axillary lumps. These were mostly on the right side when not symmetrical, and were developed generally *pari passu* with the breast.

The secretion from the lumps was of three principal kinds : (*a*) granular débris, like the secretion of sebaceous follicles ; (*b*) colostrum ; (*c*) milk. It was expressed from the situation of the sebaceous follicles, as marked by the situation of the hairs. The whole surface of the lump produced secretion ; there was no centralisation. These axillary lumps seem to prove that in lying-in women the sebaceous follicles of the

* " Journal of Anatomy and Physiology," vol. xiii.
† Royal Med.-Chir. Society, April 27, 1886 ; *Lancet*, May 1.

skin are capable of producing true mammary secretions. They confirm the opinion that the breast is a highly specialised aggregation of highly specialised sebaceous follicles. The least specialised form is that here described, where the skin is merely thickened, **and the** sebaceous glands produce true mammary **secretions.** The next form is that where there is an aggregation of the ducts which open by one or more external pores. The highest rudimentary form is where an extra nipple or more is superadded to the last variety. It is also well known that nipples may be developed independently. It is far from improbable, adds Champneys, "that these 'axillary lumps' may share the pathological affections of the breast, and even be the seat of abscess." **I would go** farther and add neoplasms, for I believe the existence of these lumps may explain the occasional existence of primary sarcomatous or cancerous tumours in the axilla, or beneath its anterior fold, which otherwise are difficult of explanation.

CHAPTER III.

HYPERTROPHY AND ATROPHY OF THE BREAST.

Hypertrophy.—When a breast gland is universally enlarged it is said to be hypertrophied, and the enlargement is supposed to be due to a simple increase in the normal structures of the gland, and not to either an overgrowth of any one of its structures, or **to the** presence of a neoplasm or new growth.

The hypertrophy in the case of the breast is supposed to be similar in kind to that of other parts, in which from increase of function there is increase of nutrition, and as a result the elements of the structure grow and multiply to enable it to perform increase of work. How far this analogy is correct may be open to a question, since it is quite certain that in some so-called hypertrophied breasts there is little or no breast secretion, even when the non-hypertrophied gland has been brought into full activity by pregnancy; and, further, that a large proportion of examples of this affection is met with in single women, and some examples in male subjects.

A woman, aged 43, came under my care some twenty-five years ago with an enormous right breast, which had been steadily increasing in size for thirteen years; it was when seen at least six times the size of the left gland; it hung down heavily, and caused distress from its mere weight. Its nipple was natural. The woman married, became pregnant, and was confined naturally. She suckled her child comfortably from the left and unenlarged gland; the right so-called hypertrophied gland gave no milk, and although it enlarged somewhat during the first few weeks of lactation, it neither gave rise to trouble nor showed signs of physiological activity.

With these facts before us, the enlargement of the breast can hardly with any correctness be called a hypertrophy, since the gland structure itself gave no evidence even of its normal development, much less of any increase in its structure; and the functions of the enlarged gland, even when stimulated by a completed pregnancy and lactation, failed to come up to the point of normal secretion.

Many years ago the same point was demonstrated in the case of a young married woman, aged 25, who came to me with an immense right breast of some years' growth, and a complaint that from some mechanical cause her husband could not have complete intercourse with her. On examination I found she had a double vagina and a bifid uterus, and from these facts the sexual difficulty of which she complained was fairly explicable. I told the patient that probably a division of the vaginal septum by means of the galvanic cautery would prove of service, and whilst she was considering the point, the question of pregnancy arose, as she had missed a period. The surmise of its possibility turned out to be true, and in due time she was delivered of a boy. The enlarged breast, however, never secreted milk, though the small or normal gland gave abundance. All sexual difficulty had vanished. It is interesting to note that in this case of enlarged breast, as well as in one of supernumerary nipples previously recorded, the genital organs were abnormal.

I have before me notes of five other cases of hypertrophy of the breast which occurred in women, and of four in men. Of the five female cases, four were single and one a married woman; the latter was 33 years of age, and the right gland was affected. Of the single women, in three the ages were respectively 19, 21, and 25 years, and the left gland was the one involved; in the fourth case the girl was 17, and both glands were implicated.

One of the single women, aged 21, had a sister under my care for a large adeno-sarcomatous tumour the size of a cocoanut, which I successfully removed. In two of the five cases the catamenia were quite regular, in three this was not the case.

Of the four male subjects, one was a tailor, aged 25, and his left breast was as large as a well-grown woman's; one was a hatter, aged 31, the affection was in his left breast, which was as large as an orange; it had been growing for six years; the third case was a man, aged 20, the right breast had been increasing for three years in a painless way, and was the size of a fist; the fourth case was in a boy, aged 18, and his left breast, which had been growing for two years without pain, was the size of half a large orange.

It is worthy of note that of the ten cases recorded in which one gland was affected, in six it was the left.

The most striking example of this affection I saw in 1865 with my friend Mr. Shipman of Grantham. It was in a single woman aged 19, and both breasts were affected. They had been increasing for a year and a half, and were removed, at an interval of three months of one another, by Sir W. Fergusson. One specimen is now at the College of Surgeons (Prep. 4,739). The gland on removal measured one foot in diameter, and weighed thirteen pounds. Except in its increased size and in the corresponding size of the blood-vessels, which (in the preparation) have been injected, there is no apparent change from the normal condition, although on microscopical examination the normal glandular structure was found to be mixed with a great increase of fibrous tissue. The engraving above was taken from this patient.

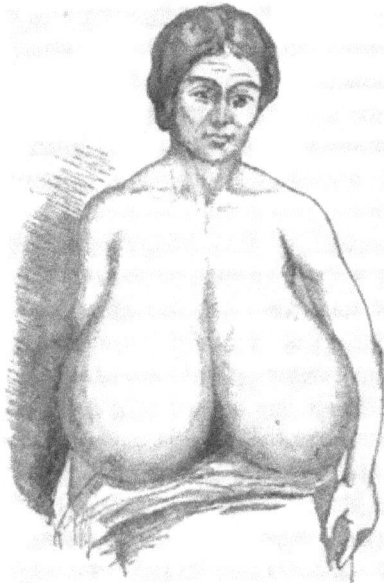

Fig. 1.—Hypertrophy of Breast.

The reader should remember that this affection may be simulated by the presence of new growths, such as lipomata, enchondromata, or adenomata. In the Guy's Hospital Reports for 1841 (vol. vi. p. 203), a case of this supposed trouble is reported, which turned out in 1843, when Mr. Stanley removed it, to be an adeno-fibroma.*

No treatment appears to have any effect in these cases except excision, which is to be performed only when the local affection is a source of serious trouble.

There is no evidence to show that this so-called hypertrophy of the breast has any connection with an inflammation. The breast in these cases, in both male and female subjects, may occasionally be more tender than the unenlarged gland, but this symptom is not always present. In the cases of inflammation of the breast in the male subject recovery generally follows treatment, and I have not seen one instance in which such enlargement became permanent.

Atrophy of the breast.—By this term is meant the diminution in size due to a wasting of the secreting structure of the breast, the result either of old age, or of some antecedent inflammatory or other change by which the nutrition of the gland has been interfered with.

When it accompanies old age, it is a physiological and consequently a natural process ; the breasts with the procreative organs wasting like other glands as soon as they have played their parts in the animal economy, and are no longer wanted for active service. When it occurs during the functional activity of these same organs, the wasting cannot be so explained, and the conclusion has to be drawn that the gland has either not been fairly formed to perform its functions when called upon, or that some pathological change has taken place in it by which its secreting power has been crippled or destroyed. The former explanation is illustrated when a breast which is apparently healthy fails to develop and secrete milk during pregnancy, and the subsequent lactating period (agalactea) ; and the latter, when a breast which

* *Vide* Museum at the College of Surgeons, Prep. 399.

c—25

has once performed its functions normally becomes the seat of inflammation, possibly of suppuration, and recovers, and then fails to respond to the physiological demand made upon it by another pregnancy.

In the wasting of old age, there may be no external or visible signs of atrophic changes, the development of fat in and about the gland rendering the actual state of the part obscure; nevertheless, on making a section of such a breast, the anatomist will often find a difficulty in distinguishing the gland, since what remains of it will consist mainly of the nipple and its ducts radiating into the fatty and fibrous tissue with which it is surrounded, the ducts often containing a thick mucus. The surgeon also will frequently be surprised to find, when removing a breast the seat of an infiltrating carcinoma, how limited the disease appears to be, although the mass removed may have appeared large; the infiltrated and contracted mammary gland presenting on section a limited tumour surrounded with fat (Plate V. Fig. 3).

In the museum of the College of Surgeons there is a specimen (Prep. 4,819 B) of extreme atrophy of the gland which I removed in 1883 from the person of a lady, aged 65, supposing it to be a carcinoma. It is described in the catalogue as follows: "An extremely atrophied breast, two and a half inches in diameter, of which the nipple is enlarged by a growth of dense fibrous tissue, both within and beneath it, and is pyramidal in shape. The other parts of the gland are of firm fibrous texture, but are less dense than the nipple." With the microscope, only fibrous tissue and a few compressed, atrophied ducts, but no cancerous growth, could be found.

Protracted or frequently repeated lactation is often followed by a rapid diminution in the size of the glands, and in exceptional instances even to their apparent complete loss. In some cases this rapid wasting is real, whilst in others the breast gland, although apparently gone, rapidly re-appears under the stimulus of a pregnancy, and becomes in due time well prepared to discharge fully its natural functions.

Some breasts, likewise, that have been the seat of an inflammation, or of a limited suppuration, are rendered permanently unfit for duty ; whilst, on the other hand, glands that have been scarred all over from abscesses and incisions, and which might be supposed to have been so injured as to have been rendered physiologically useless, re-assume, on the recurrence of their natural stimulus (pregnancy), their lactating functions; indeed, such breasts seem to secrete more fully than others that appear larger. The size of a breast, *per se,* is no real indication as to its secreting power, small glands frequently yielding more milk, and more readily, than large ones.

When tumours develop in a breast, its secreting functions may not be much interfered with, for I have known women with adenomata suckle freely, and one with a cancerous breast nurse as well on the affected as on the sound side.

CHAPTER IV.

INFLAMMATION OF THE BREAST.

THIS affection occurs at all periods of life, and in every condition of the organ ; that is, it may be met with even in male or female infants soon after birth ; in boys and girls at the age of puberty ; in women who are pregnant or lactating ; as well as in any subject male or female, who may have been locally injured.

Traumatism, under all circumstances, has a powerful influence in exciting the affection, and the **more** so **when** the gland is in a condition of functional activity. **In** exceptional cases the affection may occur in **men or** women without any assignable cause.

The inflammatory action may be **acute or chronic.** It may terminate in resolution, or in the formation of an abscess. It may involve either a single lobe of the gland, or the whole organ.· At times the inflammation will involve rather the periglandular areolar tissue **than the** gland tissue itself ; on still **rarer** occasions it will affect the connective tissue behind the gland, and give rise to a retro- or submammary, abscess.

I propose, therefore, to consider this affection under three headings, **as it is met** with during *the period of infancy, at puberty,* and during *pregnancy and lactation.*

Inflammation of the breast as seen in infants.—This is by no means an uncommon affection, and although it *may* occur without any assignable cause, it is, in the majority of cases, due to the rough manipulations of nurses who have not wholly escaped the unenlightened influence of the "Gamp" school, and think it right **either to** "rub away the milk," **or** "break the nipple strings " **of infants three or four days** old, a **practice** which **has only to be alluded to to be** condemned. It should be **well** known **that the breasts of** infants of both sexes during the first few **days** of life not infrequently contain a secretion of milk somewhat similar in character to that of **the** mother ; that is, it is made up of fluid containing colostric bodies and milk globules in different proportions. The breasts under these circumstances become full, and, **as a** consequence, tender, and at times a milky fluid escapes from the nipple. The glands **are** fullest from the fourth to the seventh day, and after that, if left alone, they gradually empty. **In** exceptional **cases** they may increase in size, inflame, and suppurate; and, should they be roughly manipulated, squeezed, or rubbed, these changes are almost sure to follow. The majority of cases of irritable or inflamed **breasts in the** infant, if rightly treated by soothing **warm sedative** applications, **such as the subacetate of lead and** opium lotion, and protected by cotton **wool from** external pressure, generally do well. The cases **that** suppurate come to the surgeon, and in these the breasts have usually been roughly treated. When they suppurate they must be opened, drained, irrigated by some iodine or other antiseptic lotion, and treated as an ordinary abscess.

In 1864 I saw **a** female **infant**, aged **two weeks,** with inflammation of **both** mammary glands, and **the** inflammation **had** clearly followed rough manipulation. **The** glands were much swollen, and the nipples **were** quite depressed in the centre of the swellings. **One** breast (the right) suppurated, the left recovered by resolution. The nipple, however, remained retracted. **I** give this case as an example of retraction of the nipple **from** early inflammatory action, believing that such **cases may** be more common than is usually believed.

An abscess may, however, occur in a child's **breast** at a later period than a month. I saw **one in 1868 in a** female, and apparently healthy, child **eight months old.** The left breast was acutely inflamed and **it had** become so a **week** before without any known **cause.** The gland suppurated, **and was opened** and treated **with a good** result.

Inflammation of the breast as met with at puberty.—At the age of puberty, even in **the** male subject, there may be signs of increased activity **in the** mammary gland, as indicated by greater fulness **of** the breast, with tenderness; of darkening of the areola and nipple.   Occasionally this increased action goes on to inflammation and suppuration.

In the female subject, remarkable structural evolutionary changes take place in association with the growth and development of the ovaries and sexual organs, and under such circumstances there are frequently present external and visible signs of this activity, which may last either for a brief period, and then disappear never to return, or may reappear with more or less force at every return of the catamenial period. Should anything occur from without, such as injury, or from within the gland during its active growing time of life, to interfere with natural evolutionary changes, inflammation may ensue, and either subside or end in suppuration. It is, however, a rare event for a breast, during the period of its development, to suppurate.

On June 2, 1862, such a case came under my care in the person of Fanny R., a healthy looking girl, aged 15, in the form of an acute abscess in the right breast. It had commenced without any known cause two weeks previously, that is, on the sixth week after her first catamenial period. The breasts had, during the whole of this time, been swollen and painful, but the left ceased to be so as the right inflamed. When I saw her the abscess was pointing, and it seemed to have involved the whole gland. It was opened by means of incisions radiating from the nipple, and treated on ordinary principles with tonics. In three weeks she was well. I have the notes of a like case which occurred in the right breast of a girl, aged 13, in whom the abscess and the appearance of menstruation were concomitant.

When the catamenia are established, and the breast gland has reached virginal developmental perfection, inflammation and suppuration are more common. It may occur as an acute or chronic affection, and possibly **the** latter is the more frequent. In the majority of cases **no** adequate cause for the trouble can be found ; in some an injury has been clearly the cause. Under both circumstances the inflammatory action more commonly attacks the whole gland **than** a single lobe. In some cases it involves **both** glands.

Illustrative cases of this affection will possibly here best prove **of value.**

Acute abscess in both breasts of a girl, aged 15.— Maria C., **a** healthy looking girl, aged 15, who had menstruated for two years, always with pain, applied **to me** in June, 1865, with an acute abscess in her right breast of six days' formation. There was no history of a blow. The abscess **was opened and** healed. Two years later, on April **22, 1867, the girl** came again with another abscess in her other breast of nine days' standing. **This** was treated by **incision** and tonics with a good result.

Acute abscess in virgin breast.—Mary S., aged 16, came before me at Guy's Hospital on May 10, **1858,** with **an abscess** in her right breast, which had been discharging for one week, after acute inflammation of three weeks' standing. The abscess apparently **involved** half the gland. No blow or known cause could be assigned for its existence. The catamenia were regular. The opening, being insufficient **for** the free discharge of pus, was enlarged, and the cavity of the abscess washed out. In one month the girl **was well.**

Chronic inflammation of both breasts in a virgin.— Emma S., a healthy looking girl, aged 15, came under my care on July 7, 1861, for chronic inflammation of both mammary glands. She had menstruated but twice, and it was after the last period, on March 29, that her breasts began to enlarge. When seen the breasts were swollen, hard, hot, and painful. There

was also some vaginal leucorrhœa. Fomentations were
applied, and iron given as a tonic, and in the course
of a few weeks the swellings had subsided.

In many cases a breast may inflame without injury or known cause, and recover by resolution without suppurating. In the one I am about to record,
the inflammation never went beyond the œdematous
stage, and then passed away. It is the only example
of the kind that I have met with. For this reason I
record it. The case explains the treatment that should
be employed under similar circumstances.

Œdema of breast (inflammatory?) ; *recovery.*—
Ellen O., aged 19, a single woman, consulted me
on July 25, 1864, for a great enlargement of her
left breast, which had been coming on for four months.
The whole gland was swollen, and the integuments
over it were thickened and tensely œdematous, but
neither red nor hotter than natural. No cause was
assigned for the trouble. Some punctures were made
into the tense skin, and much serum exuded. In one
week all swelling had disappeared, and in a month
the girl was well.

**Inflammation of the breast during preg-
nancy and lactation.**—Inflammation of the breasts
is most frequently met with in women who are, or
have been, suckling. It but rarely takes place when
the mammary glands are undergoing their physiological changes preparatory to suckling. When the
breasts are preparing to discharge their normal function, there may, in the early months of pregnancy, be
symptoms and signs of increased activity, but such
rarely lead up to inflammation, unless stimulated by
some external injury. Out of one hundred and two
consecutive cases of abscess of the breast which have
passed under my observation, seventy-nine occurred
during lactation, two during pregnancy, and twenty-one
in patients who were neither lactating nor pregnant.

Thus, roughly stated, four out of every five cases of abscess of the breast occur in lactating women. The right breast seems to be more frequently affected than the left. In my own patients this was the case in the proportion of five to three.

Birkett states that this complication of lactation occurs chiefly in women who have given birth to a first or second child. That half the cases are associated with defective nipples, and another fifth with women whose nipples are unhealthy. Thus, he adds, " we have more than two-thirds of the cases of inflammation and its results complicated with, and probably excited by, malformations or diseases of the aggregation of the excretory ducts constituting the nipple."

The affection commences usually during the early or late months of lactation. In two-thirds of my own cases it occurred during the early periods, that is, in the first two months ; and in two-thirds of these during the first month. In the remaining third it commenced during the last month. Mr. Birkett's statistics support these conclusions ; fifty-eight out of his one hundred and sixteen cases having commenced during the fourth month after parturition, eleven during the second, and forty-seven during the later periods.

It would thus appear that inflammation of the breast is mostly in the early months due to some abnormality or affection of the nipple which renders suckling difficult, if not impossible ; and in the later it is probably due either to protracted lactation, and to the exhaustion of the mother's general and of the breasts' special powers to secrete milk, or to the abrupt discontinuance of suckling from the death of the infant or other cause. In some of the cases I believe, with Mr. Ballard, that abscess in the early months is due to the searching of the child after milk before the gland is filled, in patients who have not sufficient power either to secrete milk or to resist the inflammatory process when once originated. In others, again, it is brought about by the injudicious use of a milk pump. In fact, whenever "the mamma, in its state of full expansion and perfect functional activity, becomes the subject of an interference, the result is very commonly a diffuse or nodular inflammation, and the formation of an abscess. A sudden stoppage of the milk soon after the lactation has been established is apt to produce inflammation, and the same result, or a degree of it, sometimes follows the weaning of the child after a long course of suckling. The disturbing cause, whatever it may be, acts upon the mamma when its function is at its greatest intensity, and the characteristic effect is inflammation and abscess " (Creighton).

SYMPTOMS OF INFLAMMATION OF THE BREAST.

As congestion of the blood-vessels is the earliest pathological change in the tissues of a breast that is about to inflame, so a greater *fulness*, with more or less induration of the affected part, is the earliest symptom. When the action is confined to a lobe of the gland the increased fulness will be local ; when it involves the breast as a whole it will be general. This symptom of fulness will be readily detected on comparing the sound with the affected side. With this early symptom there may or may not be *pain*, unless uneasiness of the part, aggravated by movement of the arm, local pressure, or suckling can be called pain. But as the disease progresses, pain will soon manifest itself, and its intensity will turn upon the activity of the affection, or the nervous susceptibility of the patient, but more particularly upon the tension of the inflamed lobe or gland. This tension of the inflamed part is determined by the seat and amount of effusion poured out by the inflammatory action, or by the rapidity with which it has been effused.

With these early local signs there will probably be some constitutional symptoms, as shown by increase of temperature and the general disturbance of all the functions of the body, which usually accompany every febrile action. At times, even in the very early stages of the affection, a sense of chilliness may precede the local trouble, and in rarer cases a distinct rigor may occur; but this latter symptom is more frequent in women who are suckling than in others, as in them the local congestion and inflammation of the breast is usually preceded by some local disturbance of the breast, and this local disturbance gives rise to rigor.

In lactating women *local congestion* of the breast is not uncommon. It may either subside without giving rise to any true inflammatory action, or pass on to acute suppuration, the ultimate issue of the case being much determined by the treatment it receives.

When it occurs, it is commonly called a lump, knot, or coring of the milk, and the lump or knot is essentially a lobe of the breast, which is choked with milk or its more solid constituents; this is commonly called lobal congestion. In more severe cases, all the lobes become thus affected, when the affection is called " lactic congestion."

If this local congestion of the breast does not pass on to inflammation, the skin over the gland will remain healthy, and beyond some fulness of its veins appear normal. The congested gland or lobe will, moreover, not be tender; indeed, it may tolerate manipulation freely, and find benefit from gentle friction, the friction apparently rendering the induration less marked, and relieving the congestion.

When this congested condition of the lobe or gland

passes on to inflammation, the action is, as a rule, acute, and the induration of the part primarily involved spreads to the tissues around, and gives rise either to œdema, or to a brawny infiltration of the surrounding structures. When the inflammatory action spreads forwards, the skin over the affected gland will become swollen, red, and more or less œdematous, tense, and painful, and with these local symptoms there will be the constitutional disturbance of inflammatory fever.

When the parts *behind the gland* are more involved than its coverings, there will be the same constitutional symptoms as have been described as attending the more superficial inflammation, but the local symptoms will be somewhat different, the difference being due to the presence of the inflammatory products behind instead of in front of the gland. From this cause there will be less superficial redness and swelling, but a greater prominence of the breast (*see* Plate I. Fig. 2), and a sense of deep instead of superficial effusion. The progress of the affection will also probably be slower, and there may be more pain.

Should the inflammation continue, or rather not recede, and suppuration take place, many varieties of abscess may be met with, the position of the abscess being determined by many causes. Thus, there may be a local abscess confined to a single lobe, or a diffused one involving the whole or more or less of the gland (*intraglandular abscess*). In some cases the suppuration will appear to be superficial, and confined to the connective tissue between the gland and its coverings (*superficial abscess*) ; whilst in others it will be placed behind the gland between the breast tissue and the pectoralis major, and develop as a *submammary abscess*. In rare cases the abscess forms independently of the breast, as in the cases reported farther on, but in the majority the breast gland is involved in it. The progress of the affection under these different circumstances will vary, and this variation consequently claims some little attention.

Superficial abscess.—When the **abscess is** *superficial*, the progress of the affection from the first will probably have been rapid, and although marked with the well-known local phenomena of inflammation, **it is** unusual for it to be accompanied by any severe pain, or associated with general constitutional symptoms. This mildness of symptoms is explicable by the fact that there is little tension upon the tissues, the skin readily yielding to the inflammatory effusion. Should, however, the inflammatory action involve a large proportion of the connective tissue which surrounds the mammary gland, there may be both severe local as well as general disturbance. In feeble and cachectic subjects this extension of mischief is very apt to take place, when from any cause a local inflammation has been started about the breast.

Under all circumstances a superficial abscess is prone to be followed by destruction of skin; when the suppuration is diffused, this destruction may be extensive. In patients of average power, a local inflammation terminating in suppuration may run its course, point, and discharge, rapidly, and with little loss of substance.

Intraglandular abscess is **a far** more **serious** trouble than the last variety, since its position **is** in one or more of the lobes of the gland itself; and it is to be remembered that these lobes are surrounded by a fibrous fascia, which when distended must give rise to pain. This affection is consequently attended from the first **by comparatively** severe local as well as constitutional symptoms.

The inflammatory action, moreover, since it involves deep structures, is comparatively slow in its progress, or rather, it is slow in developing the symptoms

which are usually regarded as typical of a local
inflammation; for to demonstrate these, the action
which originated in the gland has to spread through
its fibrous envelope to the surrounding tissues, and
from them to the skin. Thus it is that in this form
of inflammation many days, or even weeks, may pass
in which there may be deep mammary swelling, intense local pain, and severe constitutional disturbance
without local redness or any marked external evidence
of cutaneous implication. These superficial palpable
symptoms only show themselves when the inflammatory action has passed through the fascial envelope
of the affected lobe of the gland, and secondarily
involved its cutaneous coverings. The local pain in
the early stage of this true mammary inflammation
is often intense; it is the direct result of tension of
the tissues into which the inflammatory effusion has
been poured, and it continues until this has been re-
absorbed, or the fascial covering of the affected lobe
has yielded to the pressure from within and allowed
the pent-up fluid to infiltrate the surrounding
parts.

During the early progress of this affection the constitutional symptoms are correspondingly severe;
indeed, within two days of the earliest discovery of
an indurated and inflamed lobe of a lactating gland,
symptoms of acute pyrexia may appear, and they
are frequently associated with serious brain excitement and disturbance.

As the local affection progresses and tends towards
suppuration, the local and general symptoms alter,
although as to their intensity they rarely diminish.
The severe burning lancinating pain which ushered in
the trouble, changes into a heavy and distressing
throbbing pain, and a rigor is superadded to the
general condition of pyrexia. Indeed, from this stage
of the affection, onwards, periods of fever, rigors,

and sweating follow one another. The original seat of hardness in the breast steadily increases, until a greater part or the whole of the gland becomes equally involved ; and the breast at this stage of the trouble feels large, and irregularly indurated, **with** the skin **over it** uninvolved, and the nipple **often** retracted. **As** the abscess progresses, the skin **over the breast will** become involved, showing redness and œdema ; or **even** before, if the breast is carefully examined, the surgeon **may** detect at the original seat of induration a **soft** and painful spot, which to the experienced finger at once reveals the fact that the abscess is coming forwards and will soon burst. As soon as the abscess has burst, relief to both local and general symptoms is at once experienced, and in the most favourable cases repair sets in, and recovery is **not** far distant. To illustrate this subject I **append a** few typical cases.

Acute abscess of breast coming on during the **seventh month** *of pregnancy.*—Susan G., aged 26, the mother **of four** children, all of whom she suckled without **trouble, came to** me on January 15, 1866, with an **acute abscess in** her right breast, which had been coming two weeks. She was then pregnant seven months. The abscess was treated by incision, and rapid recovery took place. She subsequently had **a** natural labour, and could suckle with the breast which had been affected.

Abscess in the breast during the third month of pregnancy, with retracted nipple.—A woman, aged 37, the mother of one child three years old, came under **my** care on March 13, 1867, with **a** tumour in her **left** breast of six weeks' standing ; the nipple had been retracted three weeks ; there were also enlarged axil-**lary** glands ; she was pregnant three months. On April 2 the swelling, becoming fluctuating, was opened, and pus evacuated. By May 9 the abscess had healed, leaving the gland indurated ; **the preg-nancy** continued to the end.

Chronic abscess in the breast ; recovery ; two years later acute abscess in the same breast after pregnancy.—**Eliza F.**, aged 32, the mother of four children, the youngest being two years old, came to **me** in September, 1867, with a swelling in her **right** breast of eight months' standing. She had never suckled with her *right* breast, as the nipple was tender. The nipple had discharged a watery fluid up to the last month. **The** axillary glands were enlarged. The swelling was globular and semi-fluctuating.

The tumour was punctured and eight ounces **of** pus escaped ; it was then laid open and rapidly cured. On January 4, 1869, the patient returned with an abscess in the same breast **of** five weeks' standing, which had followed labour **nine months** previously, though she had not suckled. This **was** opened later on and a good recovery followed.

Abscess in breast following inability to suckle on account of deformed nipples.—Henrietta D., aged 25, came under my care in January, 1879, with an abscess in her right breast. It had followed closely upon her confinement nine weeks ago, **and** was clearly due to her inability to suckle on account of depressed and deformed nipples. She had had plenty of milk. Both breasts inflamed, but the right alone suppurated, **and** burst, discharging a pint **of** pus. The left breast recovered by resolution

Intraglandular abscess following ulceration of the nipple.—Rosie F., aged 35, the mother of four children, the youngest being thirteen weeks old, **came into** Guy's in February, 1874, with an abscess involving **the** outer half of the left breast. She had not been **able** to suckle the last child with either breast on account of sore nipples. **On** admission, the nipples were extensively ulcerated, and in the outer lobe of the left breast there was clearly an abscess, as indicated by the presence of a deep fluctuating swelling and redness with œdema of the skin over it. There was likewise much pain and general disturbance. An incision was made into the abscess, which was then washed out and drained, and in about a month the woman was well.

It seems that with her first child, born fourteen years ago, she had an abscess in her right breast, which broke in six weeks and slowly recovered. Her second child she nursed at both breasts. With the third the left breast suppurated.

Abscess in both breasts consecutive, retracted nipples as a result.—Rachel B., aged 28, a married *childless* woman, came under my care on November 10, 1864, with a globular swelling in the centre of her *right* breast, and a retracted nipple. The swelling had been coming on a week, and the nipple had become retracted during that period. The left nipple was normal. In a few days fluctuation was discovered, and the abscess was opened. Three months later the patient came again with the same condition of her left or opposite breast, and a retracted nipple. This likewise suppurated, and was treated. Both nipples were however retracted subsequently.

Abscess in the virgin breast of a woman, associated with retracted nipple.—Charlotte C., aged 25, a single woman, came under my care on May 5, 1885, with a swelling three inches by two and a half in the centre of her left breast, which had been coming for three weeks. It was hard and tender to the touch. The skin over it was natural. Nipple retracted, this retraction began with the swelling and shooting pain ; no enlarged axillary glands. The tumour was supposed to be inflammatory, cold was applied by means of a Leiter's metallic coil for several

weeks with advantage, the swelling diminishing ; **as**
soon as it was given up the swelling, however, increased, and an abscess formed, which was opened,
drained, and cured.

*Nodular tumour (inflammatory) **in** centre of
breast simulating carcinoma, following lactation, which
disappeared under treatment.*—Phœbe **T.,** aged 30, a
married woman, the mother **of one child,** now fourteen months old, came to me on October 21, **1861,**
with **a** central tumour in her left breast which had
been pronounced to be cancerous. It had been coming **for five months, and** had appeared three months
after giving up suckling. It was not very painful, **but**
was nodular. The nipple was natural. There **was**
no redness, and but little pain. It was treated as one
of inflammatory origin, and strapped up. Tonics
were also given. In the course of about two months
the swelling disappeared.

Submammary abscess is, as **a** rule, **still**
slower in its progress than the intraglandular ; although, in exceptional cases, it **may be** comparatively
acute. It is characterised by a remarkable projection
of the breast forwards from its thoracic attachments
(Plate I. Fig. 2), and on manipulation, and **more par-**
ticularly on applying pressure backwards **upon the**
gland, **a** sense of elasticity and fluctuation **will be**
detected, which is most characteristic.

The skin covering the breast **at** this time may be
natural, and there will probably **be no** change in the
appearance of the nipple. Pain **will** rarely be severe
or constitutional symptoms serious. As the **abscess**
spreads, **and by** making its way to the surface involves
the integument, both local pain and **constitutional**
symptoms become aggravated. The **periphery of the**
gland will probably be the seat of pointing, and as a
rule, the area of redness will be extensive ; it may at
first show itself in two or more local red patches,

but these areas of redness will soon unite, and perchance spread in all directions ; after a time the skin will give way by ulceration and sloughing, and the abscess discharge itself. The orifices of discharge are generally multiple. In exceptional cases the abscess discharges through the gland, when the orifice is placed near the sternal side of the nipple where the gland is the thinnest.

Acute submammary abscess after delivery ; death from lobar pneumonia.—Alice P., aged 22, was admitted into Guy's Hospital on February 11, 1875, with a submammary abscess of her left breast. It had come on after her confinement six weeks previously. The left breast was not apparently larger than the right, although it stood out from the chest in a very prominent way ; deep fluctuation beneath the breast was to be felt. An incision was made into the abscess at the lower margin of the gland and much pus evacuated ; in a few days a second incision was made higher up. Chest symptoms soon appeared, with great elevation of temperature, and the patient sank. At the post-mortem it was found that the woman had died from lobar pneumonia of the left lung ; that a large abscess cavity existed between the left breast and the pectoral muscle ; and that the breast itself was quite healthy. All the other viscera were healthy.

When this affection has been allowed to run its natural course numerous sinuses are often present, their orifices being situated around the margin of the gland.

Sinuses in both breasts following submammary abscesses two and a half years previously, treated by binding the arms to the sides.—Mary M., aged 43, the mother of five children, the youngest being four years old, came under my care on February 21, 1861, with both breasts riddled with sinuses discharging pus, the

orifices of the majority of the sinuses being at the
periphery of the breasts. These had existed for two
and a half years, and had followed acute suppuration
of both glands after prolonged suckling. No treatment had been successful. The sinuses were syringed
daily, and the breasts strapped. Tonics were also
given. Improvement followed, but no cure, until the
arms were bound to the sides to prevent muscular
movement, after which a rapid recovery ensued.

Treatment of inflammation of the breast.
—Inflammation of the breast is to be treated upon
the same principles as inflammation of any other
organ ; the object of the surgeon being, if possible,
so to check the action in its early stage as to bring
about a recovery by resolution, and where this desirable end is not to be secured, to limit the amount
of mischief in the gland, and to prevent, when suppuration has taken place, more damage to the gland
and surrounding structures than is unavoidable.

It must, however, always be remembered that
the breast gland may inflame under very different
physiological conditions, and that the treatment of
an inflamed gland in a state of physiological inactivity
must differ in a degree from the treatment of one
either preparing for lactation, as in pregnancy, or
in full physiological action, as during suckling.

The puerperal or non-puerperal condition is a consideration which should therefore ever weigh with the
careful surgeon, and in the consideration of its treatment it consequently seems to be wise to divide
cases of inflammation of the breast into two classes.

The *first* including such as occur in children and
women quite independently of the puerperal state.

The *second* those that are associated with, or
follow, either pregnancy or suckling.

In infants.—Inflammation of the breast in the
non-puerperal state, as it occurs in infants, should

not give **rise** to any serious trouble; and it may
probably be said with truth, that it would hardly ever
occur, if the breast had not been squeezed, pressed,
or damaged, by the fingers of rough nurses who **are
imbued** with wrong impressions.

Should, however, any inflammatory **action show
itself, the** application of a pad of Gamgee tissue or
absorbent cotton dipped in *warm* lead lotion, will,
as a rule, **cause it** to disappear without suppuration.
To prevent ·**a** recurrence of the action, the breast
should be protected with cotton wool, and due attention paid to **the** dresses, that they should not press
or irritate the part.

When **suppuration has** taken place **the abscess**
should be opened **as** soon as it has formed, the puncture or incision being made *in a line radiating from
the nipple.* The abscess cavity should be well washed
out with iodine water, that is, a lotion composed of
about two drachms of the tincture or **one** of **the**
liquor iodi **to a** pint of water; and some simple dressing should be subsequently applied, such as a piece
of folded lint saturated with **a** mixture of terebene
(one part) and olive **oil** (four parts); **and** bound over all
a small sheet of Gamgee tissue. With such treatment
the abscess will heal. When burrowing **has taken**
place, and the contents of the abscess do not pass away
readily, the cavity should be washed out daily with the
iodine water; in exceptional cases **a** drainage tube
may be required.

In girls.—When inflammation attacks the **breasts
of** girls **at the age** of puberty, when they **are in a**
state of evolu**tion,** the local application of **warm**
lead lotion, with **or without** the addition of opium
in solution, in the proportion of five **grains** of the
extract to an ounce of lead lotion, **will** always be
beneficial; when the action is high, saline purgation
is useful, and when the patient **is** weak, tonics

may be called for, quinine being the best. As the
acute stage subsides, the local application of the
extract of belladonna diluted with glycerine, or
rubbed down with vaseline or lard in the proportion
of ʒi to an ounce, soothes the pain, and expedites
recovery. In painful breasts, in which inflammation
may be expected, this local application of belladonna
may be all that is called for. In all cases of inflammation the arm is to be fixed to the side, to check
action of the pectoral muscle upon which the breast
rests, and the inflamed mammary gland should be
supported. In severe cases, where suppuration is
threatened, the horizontal position should be main-
tained.

In women.—Inflammation, as an acute affection, when it attacks the breast of women who are
neither pregnant nor lactating is usually the result of
injury. It occurs, however, without any such cause.
It is, as a rule, periglandular; when intraglandular it
is usually chronic. It is to be recognised by the same
symptoms as have been described above, and it should
be treated on like principles. When it first appears, it
may however be checked by the local use of cold,
applied as an ice-bag, or by what is better, Leiter's
metallic coil, a method of treatment which is of great
value in the treatment of all local inflammations, but
which is not applicable in the case of pregnancy and
lactation. Should suppuration occur, it is to be
dealt with as an ordinary abscess ; but the gland not
being in an active physiological condition is rarely
the seat of suppuration.

In pregnant or lactating women.—Inflammation of the breast in the pregnant or puerperal
woman is always a more serious affection than when
the same trouble affects a woman differently situated ;
for women in the conditions alluded to are particularly
sensitive, and are open to impressions to which others,

who are neither pregnant nor lactating, are not exposed. Inflammation, under these circumstances, attacks a gland which is physiologically active, and consequently more prone to become the seat of inflammation under the influence of any traumatism or disturbance of its functions.

The earliest symptom of inflammation of a breast physiologically active is local induration with more or less pain ; and this induration indicates a lobular congestion of the gland ; indeed, this stage may be described as the " congestive." There will not be any redness, œdema of the skin, or constitutional disturbance of any importance : gentle manipulation of the gland will not be resented.

In the *pregnant* woman this " congestive" stage, if not due to injury, must be explained by some local disturbance to the evolution of the gland, giving rise to congestion of the vessels and blood stasis, which, if not relieved, will pass on to active inflammation. To relieve it, the horizontal posture, as a means of helping the circulation through the breast, support of the gland by bandage, and gentle friction with pressure of the gland towards its nipple, are of great value ; indeed, by these means alone, aided or not by some saline purgative, the symptoms will often subside and leave no mark behind.

In the *lactating* woman, although the same treatment may be beneficial, the surgeon should examine with care the condition of the nipples, to see if there is any obstruction to any of the ducts, or irritation about their orifices, and at the same time he should examine the lobules of the breast to see if any are choked with milk, in order that their distension may be relieved, either by the manipulation already described or by some mechanical means. When suckling can be allowed, gentle friction in addition may be sufficient ; where this is impossible, some appliance for drawing

off the milk should be carefully used. **The breast
pump I** look upon with no favour, **since I am** convinced I have **seen** much harm follow **its** use **by** injudicious hands. When it is employ**ed, it** must be
used by the surgeon, and not left to a nurse, unless
she be both skilful and judicious. Where the nurse is
entrusted **with the duty of drawing the breast,** the
least dangerous **instrument is a tube with a mouth-**
piece for the nurse, or, **indeed, the** patient, to draw,
fastened to a shield to press over **the** nipple, with or
without a small reservoir between **to** hold the milk.
The patient is hardly likely to hurt herself **by** suction,
and the withdrawal of but **a** few drachms of milk
from **an** over-distended lobule will give relief **to pain,**
as well as to the congestion, and tend **to check the**
inflammatory action. The application **of pressure to**
the breast by means of strapping, at **this stage of the**
trouble is often beneficial. The **nipple should be**
left uncovered, as suckling may be **useful.**

When there are external signs **of** inflammation
such as redness ; or constitutional symptoms, such as
pyrexia, to indicate **true** inflammatory action, the
application of the lead **and** opium, or lead and belladonna lotion, applied warm, gives great comfort and
does good. The patient should under these circumstances maintain the horizontal posture, **and the**
breast should be carefully **supported,** either **by**
cushions or bandages; **one** of the **former** placed between
the arm and chest **on the** affected side is probably the
most convenient. **As the arm is** not to be used, it is
well therefore, **at the same** time, to bind it to the side
with the cushion **in position.**

Saline aperients **are often** of value, but they **must**
be judiciously ordered, since they have **a lowering
tendency,** and patients with inflamed breasts **are
rarely of** the strongest build ; indeed, **with** them, as a
rule, tonics are indicated, such as quinine, bark,

mineral acids, or even iron. When the inflammation occurs towards the end of suckling this is always the case; when it takes place in the early months, as a result of sore or malformed nipples in milking mothers, purging is of great value. When the external symptoms of inflammation are marked, the constitutional symptoms acute, the powers of the patient good, and the causes of inflammation in the nipple have made suckling impossible, the local application of from twelve to twenty leeches is often of striking benefit; they should not, however, be placed over the seat of redness, but near the margins of the gland on the affected side; by this means the vessels supplying the gland are the better relieved.

During the whole course of treatment the breast must be supported, and where pressure can be tolerated by means of strapping it should be applied.

In every case of inflammation of the breast in a lactating woman, its probable cause should be made out, and treatment based upon it. When the action is due to retained secretion, the primary indications for relieving the congested breast of milk are clear; and the **secondary,** for subduing the inflammation which is evidently the result of over-distension, are not doubtful. When it occurs towards the end of lactation, and is too probably the result of over-suckling, associated with general weakness of the body and want of power in the gland to do the work demanded of it, the indications are likewise clear; to stop the cause of trouble by prohibiting suckling, and by medicine, food, and hygienic means to improve the condition of the patient, to enable her by natural power to arrest the disease. When the inflammation is brought about by retained secretion, active measures may be taken with a good prospect of success; when from over-suckling and debility, the conditions under which the inflammatory action starts are most unfavourable,

since the subjects of the trouble, and the glands
themselves, from over-nursing and exhaustion, show
little resistance to inflammatory changes, and time is
wanted for the remedies to take effect. Under such
circumstances suppuration is almost sure to ensue.

Treatment of abscess of the breast.—
When suppuration has taken place in a breast, as elsewhere, the soundest surgical treatment is to evacuate
the pent up pus as soon as possible ; by so doing the
extension of the trouble is prevented, much pain is
saved, and the course of the affection is curtailed.
The only exception to this rule is where the suppuration is subcutaneous and very limited, and where the
surgeon cannot determine the right point for puncture.

To leave a large *subcutaneous* abscess to break, is
to allow large portions of skin to become undermined
and consequently to die.   To let a true *intraglandular* abscess take its course and find its way to the
surface, is to waste time, cause much needless pain to
the patient, and postpone recovery, and the same may
be said of the *submammary* abscess, for in it, if the
abscess is left alone, there must of necessity be burrowing of matter in all directions, and as a result the
formation of sinuses, and much unnecessary damage
to the breast and parts around.   When the abscess is
opened, the incision into it should be free, and it
should without exception be made in a line radiating
from the nipple.   By following this line the risks of
interfering with the ducts of the gland, and of thus
adding to the chances of the gland becoming useless,
are greatly diminished, and a reasonable hope may be
entertained even in the worst cases that the gland
may subsequently be able to resume its functions.

When the abscess has been opened, and it is to be
assumed freely, so that its contents may readily escape, the abscess cavity should be irrigated with some
antiseptic lotion, such as that of iodine, and the walls

of the abscess allowed to collapse; pressure should then be applied by means of strapping to keep the walls of the abscess in apposition, and the breast supported; the line of incision should be kept uncovered, and carefully guarded by some antiseptic dressing, **to prevent** decomposition, such **as** strips **of** iodoform gauze dipped in a mixture of terebene one part, **olive** oil three parts; or carbolic oil one in eighty; **and a small sheet** of Gamgee tissue, or a wad of absorbent **cotton should** be fixed over all to absorb the discharge. **In some** cases, when the pus drains from a deep cavity, **a** drainage tube may be employed. The incision into **the** abscess should always be so arranged as to assist the drainage of the **a**bscess, and the cavity of the abscess should be washed daily with iodine water, or some other antiseptic solution, either of carbolic acid, 1 in 80, or boracic acid lotion, ten grains to the ounce.

In a submammary abscess an incision should be made as soon as a diagnosis of abscess has been formed. The opening should also if possible **be** made at the lower or outer border of the breast, and a drainage tube introduced after the cavity has been well irrigated. The breast should likewise be steadily fixed, by strapping and bandage, to the thorax, so that its movement may **be** prevented and the walls of the abscess kept firmly together. In some cases of submammary abscess, or of sinuses following abscess, it is **necessary, so** as to get a good result, to prevent the **movement of** the pectoral muscle by binding the arm to the side. **In** a case already quoted the benefit of this treatment **was** well exemplified, **but in** the following it was more striking.

Sinuses about breasts following deep-seated suppuration three and a half years previously, in a young single woman.—Miss M. U., aged 23, consulted me in 1875 for **sinuses** which riddled both breasts, and likewise the submammary tissues. They had existed on the right side for three and a half years, and on the **left** for three years. They had followed acute **in**flammation which had come on without any **known** cause. Every kind of treatment had been employed without success. I simply strapped up the breasts, leaving the orifices of the sinuses open, dressed **the** wounds with terebene oil, and bound the arms to **the** sides to prevent movement of the pectoral **muscles.** In one month the patient was well.

Tonics are almost always required in all these **cases,** and of these quinine is the best. Iron when it **can** be borne is also **of** value. Good food, with stimu**lants** carefully administered, and good air are likewise essential. When the abscess is quite local, and **the** general condition of the patient good, nursing may **be** continued so as to keep the breast free from fulness, although it is not to be recommended ; but when **the** suppuration involves more than a **lobe, it** must be forbidden as much for the infant's **as for the** mother's sake.

When *sinuses* are **left** after suppuration, they may generally be healed by the careful application of well-applied pressure by means of strapping to ensure immobility of the parts about them ; when they are of some standing, the injection of a lotion of chloride of zinc of from 5 to 10 **grains to** the **ounce** is most beneficial ; when they are superficial they should **be** laid open, and treated as open wounds.

When *milk fistulæ* are present, that is, when the sinuses following an abscess discharge milk, the same kind of treatment is to be adopted as has been **just** described for sinuses ; their cure is, however, **always** troublesome ; indeed, it is not to be looked **for until** the breast has ceased to be a secreting **organ.** When the milk has "dried up," the fistulæ **will probably** heal up with it, but not before.

The induration which is left after inflammation of

the breast will slowly subside **as** the health of the
patient improves, and even after the most extensive
suppuration there is a reasonable hope that the breast
may so recover its normal condition as to **allow** of
suckling in a later pregnancy.   Should **there be** any
malformation, retraction, or disease of the nipple, this
result is not however to be **looked for,** but under
other circumstances it may be anticipated.

To guard against congestion and inflammation **of**
the breasts of pregnant women, who from **local or
general** bodily condition cannot or ought not **to**
nurse, pressure is by far **the** most reliable means.
This may be applied directly after labour in the form
of **a** chest bandage, or by means of strapping, well
adjusted ; the pressure mechanically preventing the
flow of blood into the breast which of necessity pre-
cedes secretion.

When the stage for pressure has passed, and **that**
of tenderness and congestion reached, the local use **of**
belladonna, diluted with equal parts of vaseline **or**
glycerine, is of great value.   Saline purgatives administered with discretion help towards the desired
end.

CHAPTER V.

CHRONIC ABSCESS OF THE BREAST.

In a clinical exposition of diseases of the breast,
it is clear that chronic abscess of the gland should
be considered apart from the more acute forms
of inflammation ; since examples of this affection
differ much from the more acute troubles **in** their
clinical history and symptoms, and are frequently

mistaken for new growths, such as cancer. They are met with in women **at all** periods of life, and attack the single as frequently as the married, and the sterile as often as the prolific. They follow at times **some** blow **or** injury, and are occasionally associated **with** lactation. More frequently, however, they form **without** any known exciting **cause, and are very** insidious.

The symptoms are **much those of chronic lobular** inflammation, although, if possible, less **well marked.** They are chiefly those which are attached to a local swelling, slowly and insidiously increasing with little **or no marked** pain. The swelling from the first seems to **be part of** the breast, **and to** involve it as an infiltration. **In the** early stage no other symptoms exist. Later on, there may or may not be œdema and external signs of inflammation ; but these points will best be illustrated **by cases.**

1. *Chronic abscess in the centre of the breast of a married childless woman following a blow ; nipple retracted.*—Sarah M., aged 25, a married *childless* woman, applied to me on December 31, 1863, for a swelling which occupied the centre of her left breast. It had **been** slowly **coming,** and had appeared soon after a **blow,** which she had sustained eight months previously. There had been little or no pain attending its formation. When seen the breast was universally enlarged, and formed a flattened globular swelling, in the centre of which was a retracted nipple. The nipple before the swelling appeared had been quite natural. The **tumour was very** hard and slightly tender on manipulation. No sense of fluctuation could be detected in it. A needle was introduced into the swelling for diagnostic purposes, and when pus exuded, **a** free incision was made into the gland, and **several ounces of** greenish pus evacuated. A **speedy cure** ensued.

2. *Chronic abscess in the breast of a single woman, with retracted nipple, simulating cancer.*—Ellen A., aged

56, *single*, applied to me in 1865 for a swelling which involved the whole of the left breast, associated with a very retracted nipple. The breast was **very** hard, and the trouble had been coming for **seven** months. The axillary glands were normal. The case looked **like one** of cancer. In two weeks, however, **some signs of** fluctuation were felt, and on the introduction **of** a needle for diagnostic purposes, pus escaped. The abscess was, therefore, opened, and a cure ensued. The retraction of the nipple had been congenital.

3. *Chronic abscess in the left breast of* **a** *single woman.*—Jane E., aged 32, a *single* woman, came to me **in** 1865, with a large abscess, the size of a cocoanut, in her left breast. It had been coming for five months without any known cause, and with hardly any other local symptom than swelling. I tapped it, and drew off sixteen ounces of pus. **I then** incised and drained it, when a rapid recovery ensued.

4. *Chronic abscess in the centre of the breast, with* **re-***tracted nipple, following pregnancy.*—Mary R., aged **41**, a married woman, the mother of nine children, all of which **she** suckled except the last, came under my care on August 25, 1864, with a large central indurated swelling, occupying the right breast, and **a very** retracted and depressed nipple. The swelling had **been** gradually coming for nine months during her **pregnancy**, and during this time the nipple retracted. **She** was confined at the natural term, and did well. She did **not wean** the child. When seen the skin **over the swelling was** adherent to the gland, a week **later it** was somewhat redder than normal, and in two weeks some evidence of pointing was discovered near the nipple. An incision was consequently made deeply into the gland, evacuating pus, and a rapid recovery followed.

5. *Chronic abscess in the breast, unassociated with lactation, discharging through the nipple.*—Harriet B.,

aged 40, the mother of four children, the youngest being
eleven years old, came to me in June, 1865, with an
abscess in the centre of her right breast, which had
appeared without any known cause about ten months
previously, and had discharged from the nipple three
weeks. The breast was large and tender, and pus
flowed freely from the nipple. Tonics were given, and
pressure was applied to the breast by strapping. In
one month all discharge ceased, and some thickening
of the breast alone remained, which eventually subsided.

6. *Chronic abscess in the breast of a married woman, aged 64, without known cause.*—Hannah K., aged
64, a married childless woman, consulted me on Oct. 14,
1859, for an abscess in her right breast, which had
been discharging from the nipple, and also from an
orifice below the areola for two weeks. A swelling
had existed in the breast for about three months previously. It began as a swelling that was not very
painful, which slowly increased, and then discharged.
It was cured by a free incision and drainage, aided by
tonics. The case was thought to have been one of
cancer, and on that account was sent to me.

7. *Chronic abscess of the breast, of four months'
standing, in a single woman, mistaken for cancer ;
excision of gland ; recovery.*—A. S., aged 50, a *single*
woman, came into Guy's Hospital with a central
tumour of the right breast, which had been coming
painlessly for four months, and was associated with a
retracted nipple. The skin over the breast was
healthy, and the lymphatic glands were not enlarged.
The surgeon under whom the case was admitted
took it to be one of cancer, and consequently removed the breast. Its true nature was then discovered.

8. *Abscess in the breast with retraction of the nipple,
puckering of the skin, and swelling, in a single woman ;
simulating cancer.*—Leonore W., aged 39, a single

woman, came under my care on November 22, 1866, with a swelling the size of a large walnut in the upper part of the left breast, which had been coming for months, with a nipple which was completely retracted, and with the skin over the tumour puckered and fixed to the parts beneath. The swelling was hard, and situated in the gland itself. These symptoms suggested a carcinoma. On going into the history of the case, it was elicited that the nipple had been retracted from birth, and that the puckering and adhesion of the skin were due to the cicatrix of an old abscess which she had had twelve years previously. In the course of a few weeks the tumour softened, and its true inflammatory nature was revealed. It was opened, irrigated, and cured.

9. *Chronic abscess in the breast of a pregnant woman with retracted nipple, and enlarged axillary glands; recovery.*—Mrs. R., aged 37, a married woman, with one child three years of age, came under my care, in May, 1867, with a swelling which occupied the centre of her left breast, of six weeks' standing, and which had come on without any known cause. The nipple for three weeks had been retracting, and when seen the axillary glands were enlarged. The skin over the breast was healthy, and manipulation gave rise to some pain; no fluctuation could be felt in the swelling. The diagnosis of the case was obscure. A month later it came out that the woman was pregnant about three months. The tumour then enlarged, and was indistinctly fluctuating; an exploratory puncture was consequently made into the swelling, and, as pus escaped, this was followed up by a free incision; a good recovery ensued.

10. *Chronic abscess involving the whole breast, with retracted nipple; incision; recovery.*—Eliza W., aged 34, came under my care on June 6, 1867, with a general infiltration and enlargement of her right breast, of nine

E—25

months' duration. The nipple was completely retracted, and the skin over the breast seemed to be
fixed to the gland beneath; to the eye the skin
looked natural. The tumour indistinctly fluctuated.
The swelling was said to have followed a blow. This
woman had been confined in October, 1865, and was
suckling in September, 1866, when she received the
blow. Tonics were given. Later on, fluctuation
having become distinct, an incision was made into
the centre of the gland, and pus evacuated, after
which recovery took place.

11. *Abscess in an inactive breast, with retracted
nipple; simulating cancer ; cured.*—Ellen D., aged 39,
a married woman, the mother of three children, the
youngest being eight years of age, came under my care
on April 1, 1867, with a swelling in the axillary lobe
of her left breast, and a retracted nipple ; the base of
the nipple being drawn towards the axilla. The swelling had been gradually coming for three or four months
without any known cause, and was the seat of a sharp
pain. It was vaguely globular and doubtfully elastic.
The axillary glands were sound. Its true nature was
uncertain, and for diagnostic purposes an incision was
suggested, but consent for this was not obtained. In
two weeks the skin had become glued to the parts
beneath, and some slight redness appeared ; an incision was consequently made into the breast, and some
ounces of pus evacuated. In three weeks the patient
was well.

12. *Chronic enlargement of the breast with retracted
nipple, simulating cancer, chronic abscess ? excision;
cure.*—Mrs. W., aged 64, consulted me in November,
1875, for some enlargement of her right breast, which
had been coming on for about four months. The nipple
was retracted, but this condition was said to have been
congenital. There had been but little pain in the
breast during its enlargement. When I saw her the

breast was generally enlarged and indurated. The
nipple was completely **retracted.** The skin over the
breast was natural, **and** there **were** no enlarged axillary
glands. I regarded the case **as** one of carcinoma, and
advised **excision.** The operation was performed, **and**
a rapid **recovery** ensued. The patient, **now nearly
twelve years** after the operation, **is** quite **well.** On
making a section of the gland through the nipple after
its removal, a quantity of yellow purulent fluid escaped
and **scattered** itself **over** everything. It had **been**
ejected from a cystic cavity, which occupied the centre
of **the** gland, and the walls of the cavity looked as if
they had broken down. With the naked **eye** examination I **was** in doubt **as to the cavity being** caused
by a softened **cancerous tumour, or to an** abscess.
Dr. Goodhart consequently **examined the growth,**
and reported as follows :

" I have examined **the breast tumour you sent to
me.** Without **venturing to be** too positive, **I** *think* it is
only inflammatory ; **at** any **rate,** all the appearances
met **with are** explicable on that hypothesis. The
breast tissue is still quite distinct and healthy looking,
but in places a large number of granulation-like cells
are found crowded into the fibrous **tissue,** normal to
the part. There is **no variety of** shape, and no **excess
or easy** displacement **(milky juice)** of cellular elements. Outside the breast **tissue** proper the **fat is**
undergoing definite change, **not** that of infiltration,
but of absorption of oil, and crystallisation of marga-
rine, etc., **leaving** behind cells which at first look
suspicious **from their** size, **but** which, I believe, are
nothing more **than** fat cells **gone into training,** and
having disposed **of their superfluous oil.**

"The naked **eye** appearances **are** quite those of a
chronic inflammatory tumour ; healthy breast swollen
out, **in** fact, not infiltrated.

" I shall be glad to know any result that may

hereafter declare itself, and in the meantime shall believe that it will not recur."

13. *Tumour in the breast of a married prolific woman, twelve years supposed to have been a cancer, which suppurated and was cured.*—In **March**, 1860, a married woman, aged 40, the mother of six children, whom she had suckled, the youngest being eight months, was **admitted** into Guy's Hospital under the care of **the late** Mr. Hilton, with a tumour the size of an egg in her right breast, which had been steadily growing for **twelve** years. It was clearly in the gland, irregularly **nodular and** hard. The skin over it was glued to the **tumour** beneath. **The** nipple **was** natural, axillary glands free. The tumour first appeared twelve years before, when the woman was twenty-eight, and had increased after each confinement and period of lactation. It was supposed **to** have been carcinoma. Whilst under observation, however, it inflamed, **and** discharged healthy pus. A free incision was subsequently made into it, and a rapid recovery followed. This tumour could hardly have been an inflammatory **one** from the first. It might have been at the beginning a galactocele or cyst which subsequently suppurated. When the notes of the case were taken by me **in** 1860, no reasonable explanation of the **case could** be found.

These cases, which are well worth studying, and **on that** account have been reported, **are** enough to indicate the different courses a chronic abscess may take ; the different symptoms to which it may give rise ; the difficulties that may be experienced in forming a true diagnosis ; and lastly, the errors of **treatment** that **may** be made.

They are enough to teach the surgeon **the necessity** of caution both in forming an opinion, and in dealing **with the case** when the diagnosis is not sure. They indicate the propriety of making an exploratory puncture or incision into any tumour whenever there may be the slightest doubt as to its nature, a lesson which the consideration of tumours themselves will help to enforce.

Inflammation and suppuration of the male breast.—The male breast may inflame and suppurate in the same way as the breast has been described as doing in the female, although examples are not common. Injury may, in some cases, be the cause, but in others no definite history of such can be obtained. Indeed, in the male as in the female, inflammation may apparently originate apart from anything of that nature. I find, however, from my notes of twelve cases, that the affection in eight was attributed to local pressure or injury. At puberty a boy's breast may fill up, and become as tender and irritable as a girl's; and then, either by treatment or without, get well, and no longer be a source of trouble. During young adult life the same affection may occur.

In other cases the inflammation may follow the congestive stage, which probably best expresses the condition of the breast called irritable, and this inflammation may end in suppuration. I have seen several examples of this.

All these cases are to be diagnosed and treated in the same way as when they occur in women.

The following brief notes of some of the cases which have passed under my care will probably best illustrate the whole subject. Many I reported in 1868.*

Irritable breasts in a boy.—Timothy W., aged 15, a compositor, came under my care in 1865, with swollen, congested, tender, and painful breasts. Both glands were alike, and the symptoms had been coming on for months. The boy's health was good, and no history of injury could be obtained. Fomentations

* *Lancet,* p. 285 ; Feb. 29th.

locally were ordered, and tonics prescribed. In one month the boy was well.

Chronic inflammation of both breasts in a man; retracted nipple.—George W., aged 31, came to me on January 19, 1866, with an inflammatory induration of both breasts, enlargement of the glands in the left axilla, and retraction of the left nipple. The affection had been coming on for six weeks without any assignable cause. Under the use of the extract of belladonna applied locally, and tonics, the pain became less, and in six weeks the left breast returned to its natural size. The right breast was more obstinate, but ultimately recovered with a retracted nipple.

Abscess of the breast after injury.—W. H., a lighterman, aged 31, came under my care on June 23, 1862, for some affection of both breasts. It had been coming on for one week, and when seen, both glands were enlarged, tender, and painful. He believes he struck his left breast with an oar three days previously. Fluctuation was detected in the left gland, and consequently an exploratory puncture was made from which pus escaped. The wound was then enlarged, and the abscess cavity irrigated and drained. Tonics were given, and in two weeks the man was well.

Large abscess in male breast; recovery.—Joseph K., aged 21, a healthy smith, came under my care on January 3, 1858, with an abscess in the right breast, the size of a large orange. It had been coming for two weeks, and had given rise to much pain. No assignable cause was given for its origin. Fluctuation was readily detected in the tumour, which was deeply placed in the gland. The skin over the gland was healthy. A free incision was made into the abscess at its lower and outer surface, and several ounces of pus evacuated. In three weeks, with tonics, he was well.

CHAPTER VI.

HOW TO EXAMINE A BREAST—FUNCTIONAL DISORDERS OF THE BREAST.

The examination of a breast.—In examining a breast with diagnostic intentions, the whole gland should be well in view, and the patient should be either lying down (the most desirable position) or reclining well back in an easy chair. The hand, made warm, if necessary, by dipping it into warm water, should then be placed flat, with its palmar surface upon the breast, and the gland manipulated gently with fingers and thumb in every part. By these means, if an isolated tumour be present, or any local infiltration of the breast or of one of its lobes exists, its presence will be detected ; for a tumour growing in the breast, or infiltrating it, would at once manifest its presence to the touch as a decided projection forward. The surgeon should *not* grasp the breast between his fingers and thumb, or raise it by his grasp from the pectoral muscle, until the former examination of the gland has been completed ; for in such an affection as the irritable mamma, the gland by such a method of examination would feel like either a round or flat tumour beneath the skin, under which circumstances the idea of tumour would be excited in the surgeon's mind, whereas when the whole gland is pressed flatly against the chest, with the palmar surface of the surgeon's fingers, no such apparent tumour or hard infiltrating mass will be perceptible. Under these local circumstances, and with the other general symptoms to which attention will be drawn, the diagnosis of this affection can usually be correctly made.

Irritable mamma and chronic thickening of the breast.—By an irritable breast is meant a gland which is hypersensitive, the skin as a rule sharing this increased sensibility. The breast, at the time when the pain is present, is also usually full; that is, when compared with the unaffected gland, it is larger, firmer to the touch, and the skin covering it is more vascular, as indicated either by redness from engorged capillaries, or by filled veins. The pain varies in different cases, and at different times in the same case. It may be so severe as to make the patient dread the slightest touch, even of her dress; and the examination of the breast by the surgeon may induce either a general suffusion of the patient's face and neck, or give a shock to her circulatory system, so as to bring about almost a state of fainting, as indicated by sudden pallor, nausea, and a feeble pulse. The pain is not always only local, it spreads in some cases up the neck, round the side, or down the back or arm.

Pressure upon the middle or anterior branches of the intercostal nerves, as they pass through the intercostal muscles to supply the gland, excites severe distress, and it may be noticed that now one, now another, branch will be involved. The seat of pain in the breast itself always corresponds to the branch of nerve affected. The pain is usually paroxysmal, and it may at one time affect a single breast, and at another both; or it may leave one and attack the other, and these alternations may even occur in the course of a day.

This hyperæsthesia is more common in its extreme form in young girls and single women under twenty-five years of age, than in others of a maturer age; whilst the thickening of the gland, and minor degrees of increased sensibility occur more frequently in single, middle-aged, or married sterile women from

twenty-five to **forty-five** years of age. In every case
of either young or middle-aged subjects, some de-
rangement of the functions of the generative organs
will be discovered on investigation, and the catamenia will be either profuse, painful, irregular, or
absent. There will probably be some leucorrhœal discharge between the periods. With this disordered
condition of the generative organs, the other functions **of** the body will be generally out of order, and
the nervous system will be excitable from pain and
want of rest. This local affection is never associated
with any of the usual external indications of what is
called inflammation. The skin, if more full of blood,
is never red as in a local inflammation ; and the temperature **of** the breast or body is not elevated. No
tumour, moreover, is ever felt in the affected breast
when the hand is placed flat upon it. The *diagnosis*
of this trouble should consequently in young women
never be difficult. In women between thirty and
forty-five years of age, where the nerve element is
less and the changes in the breast more marked, difficulties of diagnosis may be experienced. Indeed, a
surgeon in even moderate practice is not likely to pass
many weeks without being consulted **by a** patient
who has this affection, who thinks, or has been led to
think, that she is the victim of some **cancerous** or
other growth.

Treatment.—This affection requires general as
well as local treatment, since in the majority of cases
it is secondary to general causes. What irregularities
connected with the catamenia exist should consequently be attended to, and should **there** be any **sus-**
picion of indulgence in depraved habits, means must
be adopted to correct them. General **means to** im-
prove the bodily health of the patient are always called
for, and mild aperients, to keep the bowels open ;
tonics are, as a rule, useful, quinine or bark and

iron being the best; where pain is a real source of
trouble, sedatives may be combined with the tonic.
Plenty of good air and gentle exercise should be
prescribed.

The *local* treatment resolves itself into protecting
the breast from manipulation, and injury, by covering
it up with a belladonna plaister, or belladonna as an
ointment fixed on with strapping, so as to support the
breast. When pressure such as this cannot be tolerated, the local application of belladonna ointment
and cotton wool should be tried. In exceptional cases
the warm lead and opium lotion gives relief; or when
an attack of pain occurs, the application of a piece of
spongio-piline, upon which a solution of opium has
been poured, is likely to give great relief.

Functional disorders of the mamma.—
The function of the breast is to supply milk for the
newly born infant; and in girlhood the gland develops
up to a certain point *pari passu* with the pelvic
genital organs. But though growth may continue, it
is probable that development does not much advance
until the time has come when girlhood passes into
womanhood. When this period of life has been attained, the breast gland, with the genital organs,
undergoes further, although still not yet complete, development, since absolute ripeness waits on pregnancy.
As soon as this event occurs the breast may be truly
said to take the last step in the way of preparation for
its perfected functions, so that it may be ready when
the full period of pregnancy has been reached to do
the work for which it was intended.

The development of the gland, however, although
it is in a sense intermittent as regards its pace, is
carried out on the same lines as were laid down for
its physiological progress even in the embryo; so that
when the breast passes on from that of girlhood into
that of womanhood, it does so simply more rapidly

than it did before ; and when under the stimulus of
pregnancy the process of development goes on up to
its stage of perfection there is still but one continuous
process of evolution.

" The continuous and useful production of milk,"
says Creighton in his suggestive work on the breast,
" which alone constitutes the function from the physio-
logist's point of view, is not the essential or primary
activity of the organ, but *a mere prolongation of its
evolutionary force at its highest point.*"

This evolutionary force may at times act irregularly and oddly, and then there comes disease. At
one time, even when the gland's true physiological
function is naturally called for, there is no response,
and a more or less true want of milk secretion (*agalactia*)
is met with.   Under other circumstances the secretion
may be so excessive, that it either flows away (*galactorrhœa*) or the breast becomes so congested with its
own secretion as to give rise to serious trouble, to
which attention has been already drawn.   In rarer
cases, the breast of a girl, before womanhood, so far
as the pelvic genital organs and general physical development are concerned, may actually secrete milk,
as may the breasts of unmarried women.   Indeed,
cases have been recorded in which the breasts of old
women have secreted what has been described as
milk, and still rarer ones where the breasts of men
have been said to have done the same.

These anomalous cases need not claim much attention in a practical work such as this.   References to
works where they may be found will be given on
another page.

Agalactia.—When a want of secretion exists, the
fault may be either in the gland itself or due to some
fault in the general economy.   When in the gland
itself, there may be some organic imperfection in its
development, as shown by smallness in size ; some

wasting of the gland structure after either an attack of
inflammation and suppuration, or some prolonged
antecedent lactation. This latter cause, as it were, so
exhausting the secretory power of the gland as to
cause its atrophy.

On the other hand, the breast may be a large one,
and the subject of what is called hypertrophy, though
wrongly, and yet secrete no milk. (*See* Hypertrophy).

Occasionally the pressure of new growths in the
gland induces atrophy, but such a result is not common ;
for breasts, the seat of cancer, sarcoma, and adenoma,
often secrete freely. When the cause of want of milk
secretion is due to general and not local conditions, it
is from deficiency of constitutional power, temporary
or permanent, brought about by either functional or
organic disease.

Occasionally the administration of large doses of
iodine or of potassium iodide have been said to produce wasting and agalactia ; I cannot confirm this observation from my own experience.

At times the breast gland may be healthy, and
yet not quite prepared for suckling when delivery has
been accomplished, some delay in the appearance of
the secretion from some local or general reason
having taken place. Under these circumstances
time may work wonders when the functional activity
of the gland is stimulated by the child being put to
the nipple, and by the application of hot fomentations
to the gland.

When the secretion of the gland is evidently
scanty, the child should be weaned at once, for its
own as well as its mother's sake.

Galactorrhœa, or the constant flowing away of
milk secretion from one or both breasts, either during
suckling or after weaning, is a troublesome affection.
In some cases the flow is doubtless excessive, in
others such a term is hardly applicable, since the

discharge of a pint or a pint and a half of milk a day may be the quantity lost, and such is probably no more than the normal quantity of secretion in an active breast. On the other hand, cases have been reported in which the flow was enough, as **Dr.** Matthews Duncan states, to run through the bed and **over it to** the extent of many pints a day.* It may **occur in** healthy rich-blooded young women from **over-secretion,** the **unused** organ **overflowing, as it** were, while **the other is** being suckled, **the** milk secretion **being of** a good character ; **or, it** may **occur in feeble** women in whom the milk secretion is over-thin, such patients being bad nurses.

It may affect both or only one breast, the rule being that both glands are involved. That galactorrhœa should at times be unilateral, is no more curious **than** that after pregnancy milk secretion should be unilateral. The cause of the trouble is difficult to fathom ; indeed, no satisfactory cause has been assigned. **It** is probably **due** to some deficiency in the inhibitory force of the nerve supply. That it has a close connection with the uterine condition there is little doubt, **since** it as a rule soon ceases when the catamenial functions become re-established.

Treatment.—The chief object of the surgeon in the treatment of these cases is to bring about the re-establishment of the catamenia, and improve the general health of the patient. The administration of iron **in** full doses is therefore indicated even to affect the **uterus,** and hot footbaths to act as derivatives **should also** be employed. **Arsenic,** quinine, iron, strychnine, and belladonna have **all had their** advocates, but in no one of these drugs can any great confidence be placed : still, all are good in individual cases. Saline aperients are doubtless valuable, and **also some** restriction in the amount of liquid food.

* Discussion at Obstetrical Society, Feb. 16th, 1887.

Belladonna applied locally, with the pressure of a bandage or strapping, is undoubtedly of use ; and pressure alone has its value. Opium given in half-grain doses two or three times a day combined with quinine or iron, acts occasionally like a charm. Birkett advises the use of iodine with or without iron as medicine. Iodine has been advocated for years, but I have failed to witness its benefits.

CHAPTER VII.

SCROFULOUS SWELLING (INFLAMMATION) OF THE BREAST.

I HAVE adopted as a heading to this chapter a term applied by Sir A. Cooper more than forty years ago to a class of breast cases which he believed originated "in a low form of inflammation to which the term scrofulous may be applied," and which in more recent times is believed to have its origin in the deposition of tubercle. For many years this affection has not been recognised as a special one, and it has doubtless been mixed up with chronic inflammation and chronic abscesses. How far a distinction should be made between these affections, time and experience will determine : but I think it well to make the subject a distinct one in such a clinical work as the present professes to be.

Sir A. Cooper drew attention to it in the following words : " In young women," he wrote,* " who have enlargement of the cervical absorbent glands, I have sometimes, though rarely, seen tumours of a scrofulous nature form in their bosoms, confined, in most cases, to a single tumour in one breast ; but in

* "Anatomy and Diseases of the Breast," chap. viii. p. 73.   1845.

one case two existed **in one** breast, and one in the other. They are entirely unattended with pain, are distinctly circumscribed, are very smooth on their surfaces, and scarcely tender on pressure. They are very indolent, and vary with the state of the constitution, diminishing as it improves, and increasing **as** the general health is deteriorating. They produce no dangerous effects, and **do** not degenerate into malignancy. They **do** not require an operation ; and indeed, it would not be justifiable to **remove** them by the knife. But **I** have seen them removed, from an error in judgment respecting their nature, and when cut into, after their extirpation, they are found to be composed of **a** loose and curdy fibrine, very unequally organised."

Velpeau in his work on **the** Breast (1853), **as** translated by Henry (p. 110), writes, " Though possibly here, as **in** every **other** situation, tuberculous abscesses of **the** breast are, notwithstanding, rare ; " " the breast **is** sometimes the seat of abscesses which **may be** called tuberculous, because of the progress, and **especially** of the character **of** the pus which they contain." He then quotes an example in a woman, aged 40, **which he** believed to be tuberculous, **but** which would probably by **most** surgeons have passed as one **of an** ordinary chronic kind. Later on (p. 223), he **states** that " tuberculous tumours, without general tuberculosis, may chance to occur in the breast. They **are** usually single, and **have** no fixed sign or form. **Lumpy** and irregular in shape, they may present **themselves** in the centre of the gland, first, as under the **skin, but more** commonly they become developed beneath **the mamma. Supervening in** consequence of a blow, or without evident cause, they sometimes progress very slowly and indolently, and at **others,** more rapidly, and accompanied by *sub*-inflamma-**tory symptoms.** They **are composed** of hypertrophied,

lardaceous, greyish tissue, of a sort of cyst, which is
very thick in some points, and very thin in others; the
loculi of which contain either greyish flocculent pus
or free albuminous clots, or caseous or thick tuberculous matter, adherent to or combined with the neighbouring tissues; sometimes, however, they are represented by homogeneous masses, which are solid though
friable, and which by their form simulate pretty
closely encephaloid or colloid masses."

Velpeau then quotes a couple of cases, one in a
woman, aged 30, and another in a girl, aged 19, in
whom indolent, lumpy, movable tumours, situated in
a lobule of the gland, appeared; one was excised, and
found to be of the nature described; the second was
treated by incision and was cured. In the former,
lung disease co-existed, and after the removal of one
breast, tumours like those in the excised gland appeared in the other. Velpeau also pointed out that
this breast affection is frequently associated with some
deep-seated thoracic disease; under such circumstances,
the breast trouble is a consecutive one, and needs no
discussion. When it appears as a local affection,
these tumours are slow in their progress, and rarely
attain any size larger than a walnut. They make
their appearance as often as not without any known
cause, but occasionally after injury; they are frequently multiple. They occur in delicate lax young
women under 35, who suffer from catamenial irregularity, and are at times the seat of pain of a lancinating character: usually this is very slight, and the
pain is intermittent.

Dr. Léger writes on the same subject, and
reports * a case of a pale anæmic single lady, aged 30,
who, seven months before consultation, discovered an
indurated lump the size of a filbert at the lower part
of her right breast. This nodule steadily increased

- * *Bull. Soc. Med. d'Amiens,* 1878.

in size, and when seen was knotted, subdivided into lumps, and at the same time painful. The skin over the lump was slightly adherent and of a violet colour. The nipple was somewhat retracted ; axillary glands free from enlargement. The lady had chest symptoms and night sweats, and she died from phthisis.

Billroth reports the case which he described **as** one of chronic caseous matter, producing mastitis in a scrofulous girl, who was brought to him with nodules containing caseous matter, varying in size from a hazel to a walnut, in her breast. Each one was incised, and then cauterised with nitrate of silver with a good result.

Richet describes * a case, in which, at the outer part of a breast, a tumour the size of a turkey's egg existed, made up of three parts, which was elastic, and hard ; pus flowed from **two** incisions which had been **made** into it. When pressure was made upon **the** tumour, pus no longer flowed, but there exuded a sort of plastic lymph, yellowish, stringy, not resembling the fœtid discharge of cancerous tumours, but recalling rather the appearance of thickened lymph. Richet adds, that tuberculous or caseous tumours of the **breast,** alone, have the characteristics as above described. Other cases of this affection are on record, and I am sure I have seen some, but have failed to recognise them at the time as I ought. My attention was drawn to the subject by **an** able paper published in Gaillard's *Medical Journal of New York,* June 1884, and written by Dr. G. Durant of New **York,** who after a careful review of **the whole subject, sums** up as follows :

From a clinical **point of** view, tubercles **of the** breast present two varieties, perfectly distinct from each other : (1) a disseminated form ; and (2) a confluent form. In either, the onset is insidious ; the

* *Gazette des Hôpitaux,* May 13, 1880.

tubercular matter is developed slowly, and **without** producing pain, and some time elapses, usually, before the patient becomes aware of any trouble in the breast. Sometimes, but very rarely, even at **this** period, one or more enlarged lymphatic glands, under the inferior border of the pectoralis major may be found. They vary in size from that of an almond **to** that of a hen's egg; they suppurate, and a fistula remains, from which a serous pus, containing caseous clots, exudes. These openings, after existing for several months, may seemingly heal, but with the progress of the disease **in** the substance of the gland they re-**open** and again discharge.

In the **disseminated form** either there is **no** increase in the volume of the breast, or it is so slight as not to attract attention. The *skin* is not adherent to the anterior surface of the breast, the *nipple* retains its usual pouting form, and the **mamma moves** readily on the underlying tissues. Careful *palpation* detects hard masses in the substance **of** the gland. These are *disconnected*, variable in number, and the size seemingly inversely as their number. Usually they have the *size* of an almond, but may attain to that of a walnut. They have but *little mobility*, seeming to be connected with the adjacent tissues. Their surface is roughened by granulations, their consistence is firm, though not cartilaginous. Seldom do they give rise to *pain*. **They** develop very slowly, and often after they reach a certain size the growth ceases. In one case no increase in volume could be perceived, though four years had elapsed between the examinations. The termination of this **form is un-**known. Dubar * says that he is unable, **from his** re-searches, to say whether they may disappear spon-taneously in whole or in part; or whether, a subacute abscess being formed, they may thus be eliminated.

* "Tubercules de la Mamelle." 1881.

The confluent form.—While the disseminated
form may remain latent during the lifetime of the
patient, and only be recognised at the autopsy, this
is not true of the other variety. The tubercles by
their agglomeration and fusion constitute tumours,
which, on account of their increasing volume, cannot
long remain hidden. The breast may in a few days
double in **volume** from a sprouting out of **one or**
more of these nodules. All these changes may take
place without inflammation and without pain ; while
in other cases, febrile symptoms, gastric disturbances,
and severe pain, press hard upon a sick one. The
pain may be continuous and lancinating, or it may
continue several days, then disappear, then **return
anew.** Palpation tells us that the growth **is not**
equally distributed throughout the organ. It is confined to one portion of the **gland, and here an ovoid
body** of uneven surface and variable size may be circumscribed. Almost unmovable in the substance **of**
the gland itself, with which it seems connected, the
tumour does **not** appear to send prolongations back-
wards. It is unyielding to the touch, and yet, by
fixing the mamma by one hand, a sensation of fluc-
tuation may be perceived. In that portion of a
gland where the new growth is not so well developed,
only a diffused swelling exists. The lobulation of
the gland can with difficulty be recognised ; **the**
skin is normal, is readily moved over the anterior
surface of the organ, which itself is not abnormally
adherent to the deeper parts. Should a puncture or incision be made into one of the fluctuating points, a
varying quantity of a purulent liquid, containing
caseous clots, will escape, and the volume of **the**
breast will diminish. After a varying time (a month
and a half in one case) the fistulous openings close
and a slight puffiness alone remains. Sooner or later,
however, the swelling reappears, and a new incision

is made, or an opening appears, which remains
permanently. When the fistulæ are well established,
the disease has reached its full development, and the
appearance of the gland is characteristic. In that
part of the breast greatest in size a well-defined
tumour exists, and on a level with it one or more
fistulous openings. At the anterior wall of the axilla,
fistulæ with fungous ridges are found in the centres
of circumscribed tumefactions. The skin around
these openings is a dusky red, and for a slight distance is adherent to the subjacent parts. The nipple
may be retracted should the fistulous opening be in
its neighbourhood. Behind the fistula, palpation
reveals more or less extensive induration. At the
axilla the fistulæ are not deep, while in the mamma
the probe may penetrate from one to two inches.
The exploration is not painful, nor does noticeable
bleeding follow. The tissue exposed through the
opening is found to be very soft and friable. One
would suppose that this ever-extending, though slow
growth, would sooner or later destroy the whole organ.
It is not impossible, however, that a portion of the
breast may be spared. The termination of tuberculosis of the mamma by natural process is unknown.
Probably a deposit in some other organ, as the lung,
terminates the scene.

Dr. Durant draws the following conclusions :

(1) That the breasts may be the seat of tumours
similar to the pathological products found in many
other organs, to which the name of tubercle is given.
(2) Tubercle in the mammary gland is much more
common than is generally supposed. (3) That mis-
takes as to the true nature of these growths have often
been made, and ablation of the breast, from a belief
in their cancerous nature, has often resulted. (4)
That during life we may by clinical observation dis-
tinguish them from other tumours. (5) That as no

absolute necessity for the ablation of these tumours exists, we should refrain from operating, unless unequivocal signs of malignancy develop.

With all these conclusions, except the last, I am disposed to agree, but would suggest that in the well-marked examples of this affection it would be well to remove the gland, as by Dr. Durant's showing " the termination of tuberculosis of the mamma without intervention is unknown," and all pathologists are now convinced that persistent suppuration in any tubercular disease is prone to prove a centre for general infection, and that as a consequence the sooner it is removed the better.

CHAPTER VIII.

SYPHILITIC MASTITIS.

THIS affection is one which should be recognised, although it is far from common. I have seen but few examples of it. Why it should be so rare may be difficult to explain; it is probable that such cases have been mistaken for and treated as examples of cancer. Neither John Hunter, Sir A. Cooper, Brodie, Birkett, nor Gross, mention the affection. Lancereaux, in his treatise on Syphilis, tells us that the celebrated Sauvages * gave an excellent example of it. The case occurred in an unmarried woman, aged 30, who had been using, for several months, the extract of hyoscyamus. She presented, in each breast, a tumour the size of a hen's egg. Dense and knobby, this tumour caused lancinating pains, which extended at times as far as the axillary region, along a chain of

* "Nosologia Methodica." 1768.

glands equally hard and knobby. The patient had
ulcers in the mouth and vagina, resulting from
syphilis acquired ten years before. Keyser's pills,
continued for a month, caused the disappearance of
the painful tumours, and other syphilitic manifestations, which did not return.

Richet* says, "This tumour presents itself at first
with all the characters of a scirrhous tumour, and I
confess that in a case observed at the Lourcine Hospital, extirpation was on the point of being performed
when the discovery of another tumour, if not similar,
at least analogous, in the calf of the leg, induced us to
wait. The simultaneous disappearance of these two
tumours under an appropriate treatment completely
removed all doubt."

Ambrosoli † relates three cases, one in a male, two
in young women, aged 19 and 24 respectively. Both
of the latter presented, soon after the disappearance
of syphilitic exanthema, a diffused, firm, somewhat
painful swelling of the breast, without change of colour
of the skin. This swelling being associated with
induration of the axillary glands, iodide of potassium, in large doses, effected the removal of these
symptoms without leaving any trace of their existence. "I have," said Lancereaux, in quoting this
case, "seen a very similar one."

Velpeau, in his work on "Disease of the Breast," as
translated by Henry, 1853, mentions four cases, as
observed by Maisonneuve, and one by Richet.
One was a simple gummy tumour. In three others
there was at the same time ulceration of the skin over
the tumour, and circumscribed congestion of the mammary gland. In all, there existed at the same time
other syphilitic manifestations, such as gummatous
tumours on the head, ulcers on the legs. The cure

* "Traité d'Anatomie Chirurgicale." 1857.
† *Gazetta Medica Lombarda*, No. 36 ; 1864.

in all was rapidly brought about by preparations of iodine.

Studied in both sexes, syphilitic lesions of the breast differ in no way from those of other organs; they are most analogous to those of the testicle. They are essentially gummatous infiltrations of the connective tissue, covering the breast, behind it or in one or more of its lobules. In exceptional cases the whole gland may be involved.

The disease, where it attacks the subcutaneous tissue over the gland, will show itself as an ordinary gumma, and run its course in the same way. When it affects a lobe or more of the breast, its clinical symptoms are obscure. It may appear as a more or less defined tumour in one lobe, of which it seems to form a part, and the skin over the tumour will in this stage of the affection be natural. As the lump increases, the same want of other symptoms will be experienced, and the tumour may yet feel hard, and be painless. The skin over it may yet be free. The axillary glands may, however, be enlarged. Should the swelling form in the periphery of the gland, the nipple will be natural; should it do so near the centre of the gland, there may be retraction of the nipple.

As the disease progresses, the tumour will enlarge, and the skin over it will become involved, but as in a chronic inflammation, and not as in a scirrhus; it will become glued first to the parts beneath, then become of a dusky colour, later on, red, and last of all ulcerated; not first dimpled, then puckered, and subsequently infiltrated as in cancer. The tumour also, as these superficial changes appear, will become more fleshy, softer, and at last give rise to the feeling of fluctuation.

In this, its last stage, the disease has either partially or wholly softened down, and may discharge itself as an abscess, or the whole tumour may die as a mass, and slough out as may a bursa, thickened and inflamed by gummatous inflammation.

The progressive softening of a gumma in whatever gland or tissue it may be found, is a characteristic **symptom of** the affection ; for diagnostic purposes it consequently requires emphasis.

This disease, although slow in its progress **at first,** is more rapid towards its close. It mostly attacks women under thirty, although it may appear in a syphilitic at any time. Bumstead states that it may occur as a result of hereditary syphilis, and I am disposed to agree with him, since I have seen an example of disease of the testicles from this cause, and the two diseases are analogous. It can hardly be mistaken for cancer in its later stages. In its early stage the error may take place. The age of the patient in which it occurs, the presence of other manifestations of syphilis, the clinical history of the case, and above all, **the** suspicion that the disease may be gummatous, **will be** the best help in diagnosis.

Gummatous infiltration of breast ; sloughing of the whole gland.—Martha C., a married woman, aged 46, who has had no children or miscarriage, came to me on August 12, 1869, with an infiltration of the upper lobe of her left breast, which had been coming for eight **or** ten months. The skin over the swelling **was natural.** She had with this, enlargement of **the axillary** glands, and a suppurating node over **her left** frontal bone, **which** had been coming for six months. There was no pain in the tumour. Tonics with iodide of potassium, in increasing and full doses, were given with benefit.

On October 25, 1869, the breast tumour had become bossy, and presented the external features of inflammation, such as heat and **redness,** with fluctuation. On March 31 the breast had greatly enlarged, and was as large as a cocoanut. The skin over it was ulcerating, and the gland tissue through the opening

looked dead, and presented the yellow wash-leather
aspect, so characteristic of syphilitic deposit. In September, 1870, the whole mass sloughed out and fell
as a putrid mass into a basin, leaving a clean granulating **surface.** On November 10 the woman was
well.

No history of syphilis could be obtained, but **the**
case was doubtless a **true** gummatous infiltration of **the**
breast, as supported by the presence of the suppurating
frontal node.

The prognosis in this affection is favourable so far
as the life of the patient is concerned, and probably so
as far as the gland itself is considered, should the
diagnosis of its true nature be made out in its early
stage. A gumma of the breast ought to be as
amenable to treatment by the iodides of sodium,
potassium, or ammonium, as a gumma of any other
part. These drugs should be given in steadily
increasing **doses,** up to a drachm or more a day. In
exceptional cases the removal of the breast by operation may be justifiable. I have, however, never been
called upon to perform such an operation.

Constitutional treatment in by far the majority
of cases may be expected to bring about a cure.

CHAPTER IX.

TUMOURS OF THE BREAST.

Tumours of the breast may rationally be accounted
for by following out the functional aberrations of the
organ, and in proof of this we may all with advantage
study carefully the highly suggestive and valuable
work of Dr. Charles Creighton, on the "Physiology and

Pathology of the Breast" (1878), who has shown that
"the investigation of breast tumours reveals merely
the working of the physiological law of healthy
mammary activity under altered circumstances, that
various degrees of disordered function may result in
various **kinds** of tumours," and that **tumour disease of**
the breast is " essentially a disorder of function."

The breast, in passing from its "resting" or inactive
state to that of full activity, undergoes during the
entire period of pregnancy a process of "evolution"
which is characterised in its different stages by certain
cell changes within its acini, and transport of cells
without; and, **in the** return of the gland to **its**
quiescent condition on the subsidence of lactation, **a**
process of "involution" in which a parallel series of
changes acting in an inverse order is to be observed,
the functional subsidence of the gland being spread over
a shorter period of time than its gradual awakening
during pregnancy.

When the functional stimulus of the mamma is
acting at its *lowest* point at the beginning of " evolution " **or the** ending of "involution," the secretory products are large granular yellow pigmented cells, which
are found within the secreting acini, in the connective
tissue spaces outside the secreting structure, **and** likewise in the lymph sinuses of the subjacent **lymphatic
glands,** these cells being the **waste products of a feeble
degree of secretory activity ; and, if the** mammary ex-
citation were always to act at that degree of intensity,
the secretion, it may be said, would always be in the
form of large granular pigmented cells.

At the *next appreciable advance* in the intensity
of the stimulus, the product formed in the gland may
be described somewhat generally as a large granular
nuclear cell, which is nothing else than the crude epithelium of the middle period of the breast's unfolding
in which an imperfect secretory force resides.

Coming *still nearer to the full excitation*, the cellular
ingredients are fewer and the mucous production
much more abundant ; and finally, when the *stimulus
is at its height*, the mucous fluid has given place to a
fatty fluid, and whatever cellular elements the secretion contains are the well-known colostrum **cells**
which approach most nearly the perfect secreting cell.

The periodical unfolding **of** the breasts, which is
an obvious accompaniment of each pregnancy, is thus
characterised by a progressive series of immature
secretory products which necessarily run to waste.
The epithelial cells are not transformed into milk till
the time of delivery and during the period of suckling
following ; but the functional action of the breast has
been at work all through the pregnancy, and has
advanced in intensity just as the secreting structure
has advanced in its unfolding.   The various stages of
unfolding have corresponding secretory products, **be-**
coming less and less crude, and as there is a similar
series of more rapid but exactly parallel waste products in the upfolding, it is a legitimate inference to
ascribe " a special kind of secretory product to a certain degree of intensity of the glandular force."

When the breast gland is disturbed from its resting
state by a cause other than pregnancy, and in consequence of *some morbid excitation* is urged into a kind
of evolution process, the steps of its unfolding are less
orderly than in the normal evolution, **and the**
" spurious excitation " never carries the gland to the
end of its unfolding, or to the perfect degree of **its**
function.   And although the morbid excitation may
be said to correspond in its intensity to a stage of **the**
normal evolution, there is this fundamental difference,
that the corresponding stage of the normal process is
transient, giving place to a stronger force, while the
morbid process continues indefinitely at the same enfeebled level.   As a consequence, the cell that should

have been thrown off from the acinus as waste almost as soon as it was formed, remains in the place of its origin to multiply, and, with its progeny, to infest the glandular structure of the breast either as intra or extra-acinous accumulations. Indeed, according to Dr. Creighton, it is upon deviations from the physiological track such as these that the existence of a tumour depends.

Thus, "a circumscribed tumour arises at a particular part of the gland where the spurious excitation has advanced to a **certain stage** of evolution or unfolding ; in that particular region, probably a territory defined by the blood-vessels, the functional force has acted at a uniform imperfect level for a length of time, the inevitable cellular waste of the crude secretion has accumulated within the acini or around them, and the foundation of a tumour has been laid.

In the healthy action of the organ there **is a pro**vision for the disposal of the very considerable amount of cellular waste material by means of the neighbouring lymphatic glands. In passing from the secreting acini and in traversing the stroma of the gland, the **waste** cells often acquire a spindle form ; and although these cells are not always distinguishable from the connective tissue cells of the part, there is, especially in the bitch, a class of pigmented epithelial cells in which such changes of form and **position can be** clearly traced. The spindle-shaped waste products are the type of the peri-acinous cell collections in cystic or adeno-sarcoma.

So far as relates to the large nuclear cells, **the** *intra*-acinous collections of them correspond **to the** structure of medullary cancer, and the *extra*-acinous infiltrations of the same cells are a distinguishing feature of scirrhus. "The distinguishing feature of the less malignant form of tumour is that the spurious *functional* activity **comes nearer in** the degree of its

intensity to that of the perfect secretory *force*, the
transformation of the epithelium is a more real transformation, and the cellular waste is reduced, in part at
least, to the class of fibre-like or crescentic elements
that **characterise the** myxomatous and more benign
issues of the tumour process."

" The circumstance that the unknown diseased
excitation most commonly befalls the gland when it is
in the state of rest is of the first importance in ac-
counting for the formation of a tumour. Whether
the disturbance be a mechanical injury, or a sympathy
with excitement in the ovaries, or of a more general
emotional nature, it comes upon the breast in its resting state. The breast can react in no other way than
by following the somewhat slow process of its normal
evolution ; without the intermediate stages of unfolding it cannot reach the perfect degree of its functions
in which there would be immunity from danger. The
intermediate stages are necessarily associated with the
formation **of** crude cellular products ; it is at one or
other of the intermediate stages that the morbid force
delays, and the corresponding cellular secretion of the
gland thereupon assumes the character of a formative
or **tumour** process."

" **The** circumstances of the functional disturbance
are never exactly the same in any two cases, conse-
quently the respective modifications of structure, or in
other words, the structure of the respective tumours,
is never exactly the same."

" **It is** the climacteric effacement of the breast **that**
gives a peculiar character to the disease in women,
and there are well-marked structural differences in the
tumours according **as** they appear before or after that
period. Those that develop after the climacteric
years are perhaps the most common, as they are certainly the most intractable, and they have been the
real source of ambiguity in the pathology of the organ.

That ambiguity depends upon the circumstance that they occur in an organ which is gradually losing or has lost its characteristic structure." Where the normal itself is vanishing, the departures from the normal are elusive.

It seems probable, therefore, from Dr. Creighton's investigation, that the fibromata, adenomata, sarcomata, myxomata, and carcinomata of the breast have their type in a series of progressive changes which the gland undergoes in its physiological evolution. The feebler the intensity of the function, the more cancerous the disease; the higher or more advanced the evolution from the resting state, the more benign the tumour.

Under these circumstances, there is difficulty in making any definite classification of breast tumours; indeed, if we take the pathological anatomy of diseases of the breast, it is practically impossible to adopt any classification that is precise, which does not involve the writer of a systematic monograph on the subject in much repetition.

The breast, it must be remembered, is composed of a skeleton of connective tissue and furniture of gland cells, and there is no single class of breast tumours which may not for any individual case (scirrhous carcinoma excepted) show the free participation of both these elements. As one or the other predominates in any individual tumour, so will it be called sarcoma or adenoma, according to the prevailing opinion of the day, although practically all the cases are essentially of the same complex structure. Furthermore, any tumour in a tubular gland must almost of necessity be liable to the formation of cysts; its very structure presupposes it. Thus all tumours of the breast more or less are associated with cysts, some largely so. Neither the presence of a cyst in a breast tumour, nor of many cysts, has, however, much pathological meaning, although in a clinical sense it so happens that in a certain group of breast tumours which are largely and chiefly cystic, the term cysto-sarcoma has been applied. **For** the great majority of breast tumours it is **therefore** true that in any one **of them** there may be, at the same time, connective tissue, gland **acini** and the derivative of gland tissue, viz. cysts.

Some, and these chiefly foreign observers, **have** classed most tumours which are not **carcinomatous with** the sarcomata, on **the** ground that histological investigation shows that the connective tissue is the initially aggressive element, but it **is** probably a sounder and **at** any **rate it is a more** universally applicable view to regard the **type of** breast tumour as a composite **one, and therefore it seems expedient to** retain the old **term adenoma for many of them, since** to call **all these tumours sarcomata is to** place **them in a** group **which prevents one from saying** positively that **any individual case is free from the** risk of recurrence, **when clinical** experience **teaches** us in no uncertain way that the solid adeno-sarcomata of young **women are** for **the** most part free from any such risk, **and** that a large number of the cystic tumours are equally benign. On the **other hand, there** is not one of them, be it myxoma, fibroma, or sarcoma, **which may** not return should it develop under the faulty inhibition of later life. I cannot, therefore, agree, **as** I should **wish to do,** with the view of that excellent surgeon, **Dr. S. W.** Gross, of America, as expressed in **his recent work** upon the Breast (1880), and draw a fast **and clear line between** these varieties of tumours; indeed, I would **rather,** after the fullest consideration of the subject, **still** prefer to maintain the **word adenoma** as indicating **the special** gland elements in most **of** the varieties **of** neoplasm of the breast, adding **the** word fibroma to **the more** fibrous form; the **compound word,** " adeno-fibroma," **clearly** expressing the pathology of the class of tumours as well as its clinical peculiarities. The word will therefore be thus used in this work. On this matter I will, however, let Gross speak for himself :

"While it is true," he writes, "**that hyperplasia of** the glandular tissue of the breast may **be so** excessive as **to** constitute the **tumour known as** adenoma, this occurrence is so rare that **true ade-**nomata must be ranked among the most infrequent **of** neoplasms." "In all of the connective tissue tumours, the lacteal glands, although their epithe-**lium may be in a state** of irritation, generally **remain passive, and as the** growth advances they **may disappear to such** an extent as to be scarcely, if at all, recognisable. Instead, therefore, of being newly formed and predominant, the secreting elements are really merely accessory or accidental, and represent the remains of old or pre-existing glands, contained, but widely separated, in a fibromatous, sarcomatous, or myxomatous stroma" (p. 304).

I should like, moreover, to add that the diagnosis between any of the varieties of breast tumours not carcinomatous is uncertain and difficult, and whether a tumour is to be called a fibroma, adeno-fibroma, adenoma or adeno-sarcoma, can only in the majority of cases be determined by a histological examination of the growth after removal. Clinically I would therefore still divide breast tumours into two great classes, the carcinomatous, and the non-carcinomatous. dividing the latter large class into the adeno-fibromata, adeno-sarcomata, and cysto-sarcomata.

Under these headings the different tumours will therefore be considered. I shall treat of the carci-**nomata** by themselves, and likewise all the cystic tumours, with or without intracystic growths ; the lipo-**mata,** chondromata, and colloid tumours receiving their share of attention.

CHAPTER X

ADENOMATA AND ADENO-FIBROMATA : THE FIBROMATA
OF GLANDS (A SARCOMATOUS ENCAPSULED GROWTH).

UNDER the term adenoma without doubt a large variety of tumours of the breast have been grouped, this
word having included "sarcomatous tumours" of
every variety ; the "chronic mammary tumour" of
Sir A. Cooper ; the "mammary glandular **tumour**" of
Sir J. Paget; the " tumeur adenoide" of Velpeau; **the**
"corps fibreux" of Lebert, and the "adenocele" **of**
Birkett. At the present day this grouping is **not**
recognised, consequently, to place myself **in accord**
with its teaching, I shall in the present volume describe several groups of tumours of the breast.

The "fibromata" and "adenomata" of the breast
may, in a pathological point of view, be kept distinct,
yet it is not possible clinically **to** maintain this distinction ; **since in** bedside **work** it is beyond the
power of the surgeon to diagnose with **any** certainty
the one variety of growth from **the other; and** it is
only by a microscopical examination of any individual
tumour after its removal that its true pathological
position **can be made** out.

It **is well, therefore, to know** that **such a definite**
diagnosis is **of no practical** importance, and **that the**
treatment of **these** varieties of tumours is the **same.**
Indeed, in a pathological sense **there is** no **great**
difference between **the glandular and** connective
tissue groups, since **all** are made up of connective
fibrous tissue, ducts, acini, and the cæcal terminations
of the ducts in varying proportions. The growths,

G—25

as a whole, form a chain, of which the links at one
end are composed of fibrous tumours, in which the
fibrous elements predominate; and at the other end of
adenoid tumours composed of more or less typical
glandular elements, and containing even glandular
secretion; whilst the intermediate links represent
tumours of a mixed fibro-adenoid type. A connective
tissue tumour, which in non-glandular structures
would be either a fibroma, sarcoma, or myxoma, when
found in a gland, which is normally made up of gland
structure and connective tissue, becomes mixed up
with the glandular elements in very variable degrees,
so that in the breast where these varieties are best
seen we have adeno-fibroma and adeno-sarcoma.

Whilst the special characteristics of the tumour
may be determined by the changes which are found in
the connective tissue of the gland, the changes in the
glandular elements themselves are not to be ignored
or passed by as of no significance, since I hold it is
due to the presence of these gland elements, and the
changes that take place in them in morbid processes,
that the pathology of breast tumours differs from
the pathology of tumours of the ordinary connective
tissue unconnected with the gland structures.

It must be remembered that a gland is a compound
structure composed of connective and fibrous tissue,
with the special gland element in the form of cell
structure. In the breast this gland structure is
tubular with clusters of cæcal acini lined with epithelium appended to the tubes.

The breast glands, moreover, are subject to
periodical excitement, and at times to active secretion.
Under these circumstances, though in variable degrees, increased nutriment is sent to them, and with
it increased physiological force is supplied. Should
this force fail to attain its true physiological outlet and
secrete milk, or should it from some outside or inward

cause be excited at wrong times, and as a morbid
condition, it will probably tend to mischief, and in a
clinical point of view this mischief will, as a rule,
take the shape of a new growth.

When an unnatural or morbid excitation affects
any part of the gland, and as a consequence growth
appears, the result may be overgrowth of any one or
of all the elements of which the gland is composed;
under such circumstances the gland as well as the
connective tissue structures will increase, and give
rise to either the adeno-fibroma where the fibrous
tissue is in excess, to the adeno-sarcoma where the cell
structure predominates.

In the early history of gland life, the force which
the part has supplied to it, in common with the other
regions of the body, is registered in growth and
development. At a particular period, again, in its
existence, partly, no doubt, owing to calls that are
made upon it, and partly, also, to a tendency long
inherited to take on a definite action at a definite
time, the gland leaves off growth and takes on the
capability of secretion, and thus is provided an outlet
for its energy. Supposing now that, in the full tide
of this developmental activity and high nutrition, a
part of its secreting surface becomes shut off, or
injured and excited when its dormant power
cannot find equilibrium in secretion, what more
likely (if, indeed, it be not just what we should
expect) than that it should go back to its prior formative processes, and commence again to grow? Is not
this physiologically sound? At any rate, presumptive
evidence in its favour is afforded by the fact that
these adeno-fibromatous, cystic, solid, and adenomatous
tumours are most common in young unmarried
women; in other words, in those in whom secretion
is in a state of nascent activity. Give this no outlet,
and new growth will very possibly result; but, more

than this, the new growth is likely to be of such a kind as shall be a fair index of developmental tone, and of an otherwise healthy state, viz. a fair substitution for healthy gland structure. That this is true is demonstrated by the fact that in certain adenomatous tumours of the breast, encapsuled as ordinary fibromata, the new structure may secrete the normal secretion of the breast. The two following cases taken from Mr. Birkett's experience demonstrate the fact. They have been published in the Guy's Hospital Reports for 1855.

"*Case* 1. *Adenoma with ducts and secretion.*— E. A. K., a healthy married woman, when 21 years of age, and a few weeks after marriage, in 1850, discovered a small lump in her left breast. It was situated in the axillary region of the organ, and she is quite sure it never entirely dispersed. I saw her for the first time in January, 1852. She gave birth to her first child in January, 1851, which she suckled, and mostly with the affected breast, until October, when the infant died. At the time I saw her she was pregnant with her second child, and the condition of the left breast was as follows: it was very heavy, solid, and I thought I could detect fluctuation. It was about double the size of the right breast, and the enlargement seemed chiefly in the axillary region of the organ. The external appearance was simply that of inordinate size and fulness, globular, and projecting from the thorax. The nipple was displaced considerably, and although never well developed, took a direction forwards, inwards towards the middle line of the body, and upwards. Scarcely any pain was ever experienced in the part, but only an uncomfortable sensation of weight. Doubtless, the fact of her being in advanced pregnancy gave rise to considerable difficulty in arriving at an accurate diagnostication of the disease, for it was not until after parturition,

in April, 1852, that a new growth could be distinctly
felt. The second child **she** suckled some months. A
third child was born **in December,** 1853. This **she**
suckled one month, and then **it died.** Whilst suckling
the second infant, milk abscesses formed consecutively
in the left breast, which obliged her during the **later
months of** lactation **to** suckle **with** the right **gland**
only. For several months **previous** to **the** operation
for the removal of the tumour, **its** nature was very
clear. During the later months of suckling, and after
weaning, the normal gland being relaxed and flaccid,
the firm solid new growth could be grasped by the
hand, and isolated from the breast. At the same
time the skin rolled freely over the tumour, and the
lobulated surface of the growth was distinctly traceable
beneath it. A very large vein **was** also seen traversing the surface of the tumour, **and** during **lactation
the** dimensions of this vein were enormous. **It took**
a course towards the axilla, where **it** was lost. On
the 28th of February, 1854, I removed **the** tumour by
merely dividing the skin over it ; the mass being only
very loosely connected with the gland, not a lobe of
the true breast was cut, but it was left entirely in its
normal situation after the tumour which displaced it
was detached. The tumour weighed three pounds.
Very **little** hæmorrhage followed the incisions ; the
edges were adjusted by means of plaister, and the
wound healed in about a month. The tumour was
composed of lobes loosely connected together. Each
lobe could be separated into lobules ; these again
into glandules or acini, which contained the cæcal
terminations of **the** ducts. **From these** the smaller
ducts united to form larger, and they traversed in some
places the interlobular spaces. The free ends of some
of these ducts terminated loosely on the surface of a
cyst, which contained a large collection of a soft solid
material, resembling thick cream. An analysis of

this compound, made by Dr. Odling, afforded the
following results :

Fat	...	...	...	...	85·31
Water ...	...	...	...	...	11·43
Albuminous matter, casein ?...				...	1·63
Animal extractive	...	...	...		·21
Inorganic matter	...	...	...		·58

" Besides this large collection of cream in a rather
thin-walled cyst, there was not far distant from it, of
smaller size, another cyst, filled with an almost cre-
taceous material. The wall of this smaller cyst was
firm and rigid, or like a piece of thick parchment.

" *Case* 2.—The only case, except the last described,
in which I have found ducts containing milk, was
shown to me by Mr. Nathaniel Ward, who removed it
from a patient in the London Hospital. The following are the particulars of the case :

" E. D., an unmarried and healthy girl, when
23 years of age, discovered a lump in her left breast,
about three and a half years before she showed it to
Mr. Ward. The catamenia did not appear until she
was nineteen years of age, then they ceased for nearly
a year, and when they again reappeared, the general
enlargement of the breast commenced. For a few
months before the removal of the tumour, a well-defined and lobed mass, connected with the gland, was
perceptible. It was situated behind those lobes of
the gland which form its sterno-clavicular quarter ;
but with care the normal gland tissue could be distinctly detected, independently of the tumour. The
lobes of the tumour were clearly visible beneath the
skin, and a large vein traversed its surface. It was
removed in May, 1855, by Mr. Ward, without any
of the breast, and that gentleman sent me a section
of the new growth. It was one of the firmest, most
dense, and closely packed, small-lobed tumours, which

I have examined. It had an extremely irregular
external surface, and between the lobes or in the
interlobular spaces, which were united together by
loose filamentous tissue, were some large veins. These
could be traced into furrows between the irregularities on the surface of the tumour. Ducts were also
distinctly seen taking a course directly transverse to
the large veins, and apparently disposed upon the
surface of the lobes and between them, without any
very definite arrangement. Many of these ducts
contained secretion exactly resembling cream. The
follicular terminations of the ducts were very perfectly developed, and were very minute, although not
so small as those of the normal gland tissue. No
juice pervaded the mass, but in some places this
creamy secretion of the ducts was easily expressed by
the cut surface.

"The mammary glands, the nipple, and areola, were
in this instance remarkably characteristic of the
virgin state. It is, therefore, a most singular fact
that a new growth should possess the function of
a normal gland, although somewhat imperfectly, especially when those organs, which the new growth
closely resembled, had never been stimulated to the
performance of their own peculiar function.

"**We** shall now take a comparative review of these
two very interesting cases : Both the patients were
very young women, in good health, when the
tumours were first observed. One had been mar-
ried, and **was in** the early months of pregnancy
when she **discovered the** tumour. The other was
an unmarried **woman,** and a virgin, when the
development of the tumour began. In both cases
soon after the observation of the tumour, considerable difficulty **arose** in diagnosticating the dis-
ease ; in the first case, on account of the general
enlargement of the mammary gland coincident with

pregnancy; in the second, because the development
of the new growth was accompanied with general
enlargement of the entire organ. Until this general
enlargement of the organ had subsided, the tumour
was not perceptible or distinguishable by touch or
sight. When, however, in one case, lactation had
produced a rather relaxed state of the breast, and, in
the other case, the general excitation of the organ
had subsided, then, and not till then, the diagnostication of the disease was comparatively easy.

"These remarks are introduced in order to show
the very great caution which should be exercised in
pronouncing an opinion upon the nature of almost all
diseases of the breast during any active or excited
condition of the normal mammary gland. Surprise
would perhaps scarcely be excited by the existence of
the solid and fatty particles of milk, in the case in
which the function of lactation had been performed
upon three separate occasions while the tumour was
growing. But the other case demonstrates that even
a virgin may have a tumour of this nature developed
in the breast, and that the new growth may so closely
resemble the normal gland, and be so perfectly well
developed, as to secrete the fatty and oily particles of
common milk.

"In both these cases the tumours were removed
without cutting away the smallest piece of the normal
mammary gland, a proceeding highly important in a
practical point of view; indeed, in all cases of a like
kind, the breast itself should be carefully sought for
and preserved. This important fact is alluded to
by Sir Benjamin Brodie."

At times these breast growths are composed of
loosely connected fibre tissue, with but a small ad-
mixture of the glandular elements; the growth
would then be called fibro-cellular. The best example
of this kind that I have seen I now record. The

clinical history of these cases, however, and the chances
of complete immunity from future trouble after the
removal of the primary growth, are the same as in the
most benign tumour.

Fibro-cellular tumour of breast, weighing, on
removal, four pounds thirteen ounces. It had commenced during pregnancy, and had grown rapidly
during lactation. It was removed, and a good recovery
took place, the woman being well fifteen years subsequently.

Louise H., aged 34, a married woman, the mother
of five children, came under my care on March 17,
1868, with a tumour in her left breast, which had
been growing for eight months, and had made its
appearance during the sixth month of pregnancy.
She had been confined when coming under observation five months, and was suckling with the healthy
breast, but not with the affected one, although at
times milk ran from the nipple. The tumour, during
three months, had grown rapidly. When seen the
left breast formed a large tumour, which measured
twenty-five inches in circumference, and at its base
seven inches. It was quite movable over the pectoral
muscle, and the skin over it, though stretched, was
not adherent. The nipple was flattened out. The
surface of the tumour was somewhat nodulated, and
the mass felt solid and elastic, but it did not fluctuate.
The temperature of the breast was *higher* than the
healthy gland. The axillary glands were not enlarged. The tumour was excised, and weighed on
removal nearly five pounds.* " Under the microscope the tumour was found to be composed of
fibro-cellular structure in various stages of development. A certain portion was made up of wavy tissue
fully formed. A larger portion was composed of elongated cells, and fibre cells splitting up into bundles

* Guy's Hosp. Mus., Prep. 2293ᴺ.

of wavy tissue, and in this structure a large number of nuclei were observed. In addition to these elements, simple cell structures were observed in various portions of the tumour. These cells were for the most part closely packed together, so that in a section composed chiefly of wavy tissue and fibre cells, here and there an accumulation of simple spherical cells could be seen." *

It seems that on rare occasions these growths may appear in infancy; when so, they are not to be expected to show the complete glandular elements of adult life; on the contrary, they show more of elementary structures, such as the fatty and fibrous, with some tubes. In the following example these points are illustrated, the best glandular elements of the case being the presence of tubes.

Lipomatous adeno-fibroma in the breast of a male child ten months old; operation; cure.—Absalom S., aged 10 months, was admitted into Guy's on July 20, 1874, under the care of my late colleague, Mr. Cooper Forster, with a hard tumour, the size of a walnut, situated above and to the outer side of the left nipple, which moved freely upon the deeper parts, but to which the skin was adherent, but not discoloured. The mother had noticed the lump for about four months, and it had grown rapidly. The growth was excised on July 22nd, and found to be made up of fat and fibrous tissue, with some imperfectly developed breast tubes. "The tumour showed a number of tubes like breast tubes, and outside them a germinating fibrous stroma, containing oval nuclei" (Goodhart). The child did well.

Macroscopical appearances of the adenomata and adeno-fibromata.—The external physical characters of these growths after removal,

---

* Committee on Morbid Growth. Path. Soc. Trans., vol. xix. p. 388.

as well as the appearances of their surfaces on section,
are very variable, and yet in both aspects they possess
a strong family likeness.

That is, all are *encapsuled*, although the capsules
may vary much in their density; all are *lobulated*,
some but slightly, others in a very marked bossy way
(Plate I. Fig. 1).

All have, on manipulation, a firm and *fleshy feel*;
and they rarely, if ever, are so hard as a scirrhus.

Some of these tumours, although a small minority,
are intimately connected with the breast, so that
to remove them, a portion of the gland itself will
have to be taken away. Others, which form the
large majority, are so loosely attached to it, that they
may be shelled out, on the division of the capsule
which surrounds them. In a few cases the tumour
is pedunculated and appears as a kind of outgrowth
from the edge or one of the surfaces of the gland.

On section, the cut surface of every variety of
these two forms of tumour has a tendency to become
convex, and to exude a tenacious mucoid fluid. It
has, when fresh, a white, pinkish or red appearance, according to its vascularity, and presents many different
forms of lobulation, as well as degrees of density.

In some cases the surface will show either a coarse
or fine fibrous structure, and under such circumstances, little or no glandular elements will be found;
the tumour belonging to the fibromata. Whilst in
others the most striking feature will be that of lobulation. It occasionally happens that these two varieties of tumour are found in the same breast. In
examining a series of specimens of this kind, the
varieties of forms which the lobules assume is very
remarkable. Where the lobules are large, the growth
will generally be succulent; when they are small, it
will be compact. In the compact form the microscopical structure of the tumour will be that of a fibroma,

including in its concentric interlacunar bundles of
fibrous tissue the cæcal terminations of the ducts
and acini of the gland. In the more succulent
forms of growth the tumour will be more or less
furrowed and fissured, foliated or dendritic, and ducts
in the process of development will be seen mixed
with the true glandular elements. In some cases,

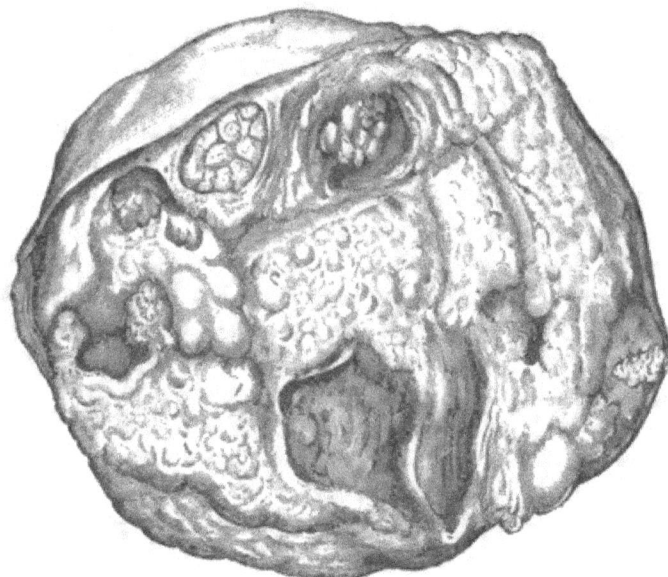

Fig. 2.—Section of an Adenoma, or Adeno-fibroma of the Breast.

again, the lobules will be compact, smooth, and apparently homogeneous; whilst in others, or even in
another part of the same tumour, the lobe will be
subdivided up, and each small lobule will be apparently, except at its base, free from connective tissue
attachments to the neighbouring lobules (Fig. 2).
Under these circumstances, it will present to the eye
much the appearance of a coarse papillomatous growth
pressed into a capsule, each papilla being composed
of the cæcal terminations of gland structure, filled

with epithelial cells, and **held together by** fibrous elements.

Histologically the epithelial elements will be normally arranged (typical) upon their special basement membrane, and although these elements may be found in excess within the acini of the gland structure, **they** will still possess **a kind of regularity in their** arrangement.

In carcinoma no such regularity will be observed. The epithelial gland elements will not only have no special relation to the cell wall, but they will be found in as great abundance outside as within the **cell,** and irregularly placed. In fact, **they** will be found to be infiltrating the whole structure. The presence **of the membrana propria** in adenoma, and its absence **in carcinoma, differentiates** the **one disease** from the other.

In many **of** these tumours, rounded cavities **or** irregularly curved spaces are visible, which contain a mucoid fluid, possibly more or less blood-stained. **These spaces** are sometimes dilated ducts, containing degenerating epithelium, and sometimes apparently spaces **due to** the distension caused by the degenerate products of the epithelium of the growth. In certain examples of the solid lobulated adeno-fibroma occupying one of the lobes of the gland, and associated **with it,** cysts will be found of all sizes, more or less occupying portions of the gland. The cysts may be present **without** as well as with intracystic growths, and these **growths may** be of different dimensions and of different structure.

Such growths **will, however,** to a certainty approach the adenomata in microscopical structure, **and** show a very **variable amount** of **connective** tissue elements between the lobules, some being dense, others **very** loosely connected, indeed, probably floating in the serous fluid of the cavity (duct?) in which it is growing.

The co-existence of these varieties of growths in
the same gland, at the same time, raises the question
as to the origin of the more solid forms, and
suggests the probability that they had originally an
intracystic origin, and had filled the cyst. This probability is likewise supported by the clinical fact, that
in these cases of solid growth there is occasionally
found during their development a serous or sanguineous discharge from the nipple, which can be
increased by pressure ; and also by the pathological
facts that in certain specimens* a tube can be introduced into the duct near the nipple, and from it
all the cysts composing the mass of cystic and
intracystic growths inflated ; while in certain cases in
which, after removal, fresh growths appear, the first
tumour will be of the more solid or fibrous variety ;
the second in time of age, cystic with loose intracystic growths ; and the third like the first,† solid.
The first and last growths being adeno-fibromata, the
second cystic adenoma. The following case of Mr.
Birkett's illustrates this point :

*Of the alternation of the so-called "chronic
mammary" tumour, and the "cysto-sarcoma" in the
same individual, and at different periods of life.*

A healthy, married, but sterile woman, was first
seen by me in June, 1851. She was forty-five years
old, strong, and in good health, although lately she had
felt weak and sinking, and had become rather thinner.
The catamenia were not regular, and at times she had
suffered with dysmenorrhœa. Four or five months
before her application to me she discovered a lump in
her right breast ; it had been preceded by a pricking
and shooting pain, and these sensations had attracted
her attention to the part. When I saw the tumour
it measured about one and a half inches square; it was

* Guy's Hosp. Mus., Prep. 2293³⁵.
† Guy's Hosp. Mus., Prep. 2299⁴ᵇ and ⁴⁶, and a drawing 402⁴³.

seated on the surface of the gland, and rather towards
its sternal border. It was very firm, movable, lobed,
and irregular on its surface, and it seemed quite detached from the tissues of the breast itself. The
woman suffered great pain upon gentle manipulation,
and even for some time afterwards. The nipple was
unaffected and the axillary lymphatic glands were
quite healthy. She said her sister had died of a cancer
in the face. Her general health at this time was a
little deranged, her appetite bad, and she complained
of weakness. Soothing applications were employed
over the tumour, and her general health being improved by the exhibition of medicines, the tumour was
removed on the 16th December, 1851, by making a
single vertical incision through the integuments. No
part of the breast itself was removed. The new
growth was enclosed in a loosely attached envelope;
it consisted of three unequally proportioned lobes,
and these of lobules which were all associated together
by loose connective tissue, in which the cæcal terminations of the ducts loaded with epithelium were
enclosed, constituted the lobules.

The wound healed favourably, although the whole
gland was extremely tender and swollen, soon after
the operation. During the whole year 1852 the
breast was usually painful at the catamenial periods,
and there appeared to be considerable local excitement
in the organ.

In November, 1853, two years after the first
operation, a new growth was discovered near the
cicatrix; the whole breast was very painful; and
in February, 1854, I removed this second growth,
together with the cicatrix and a very little adherent
gland tissue. The dissection of this new growth exposed several cysts of variable size, and which were filled
with intracystic growths, minutely and finely divided
into lobules: these were very loosely united together by

connective fibre tissue. In another part was a growth much more resembling in appearance the first growth that I removed ; it was firm, lobulated, enveloped in a fibrous capsule, and entirely united into a solid mass by connective fibre tissue. As in the first growth, so in these, the minute elements of gland tissue were distinctly visible when magnified ; and although the cæcal terminations of the ducts differed from those of the first growth in many particulars, yet they were well marked and characteristic.

The wound healed favourably, but during the summer months of 1854 she suffered great pain in the axillary half of the organ ; the whole became much swollen and indurated, and resembled the breast of a woman during the later months of pregnancy.

In October, 1854, I discovered a third growth at some distance from the cicatrix, and among the axillary lobes of the gland. This increased rapidly, and was accompanied with general swelling and induration of the breast. I removed the third growth in February, 1855. It was much larger than either of the others, was embedded in the lobes of the breast, and was firm, dense, lobulated, and united together by connective fibre tissue. In this growth the cæcal terminations of gland tissue were distinctly seen.

In July, 1855, the patient was well. The principal facts of the case are as follows : The patient when the first growth was discovered, was 45 years old. The first growth was eleven months old when removed ; between the removal of the first growth and the discovery of the second, one year and ten months elapsed ; the second growth was three months old when removed ; between the removal of the second growth and the discovery of the third, eight months elapsed ; the third growth was four months old when removed. All the growths, varying as they did in their external appearances, still exhibited, when carefully and minutely examined, those elements and structures which are found in the mammary gland itself, with the exception of ducts. All, therefore, come under the denomination " adenocele."

The first and last were, however, of the same nature as the tumour termed by Sir A. Cooper, " chronic mammary tumour." The second growth belongs to that class of growths described by many writers under the term " cysto-sarcoma."

This case, therefore, clearly demonstrates, that in one and the same individual, at different times, these new growths may alternate with each other. It likewise shows, that the second growth grew more rapidly than the first, and the third more rapidly than either of the other two ; and also, that the interval of time between the development of the second and third growth was much shorter than that between the first and second. These facts, which repeat them-selves in the natural history of recurrent new growths, are now generally recognised.

Such cases as these form as it were connecting links between the more solid varieties of fibromata and fibro-adenomata, and the cystic forms of benign breast disease, to which attention will be drawn in the chapter devoted to the subject of "cystic disease with intracystic growths."

When these fibro-adenomata are undergoing de-generating changes, they will appear to the eye of a yellow colour, and by pressure or scraping the cut surface, cheesy dead epithelial cells may be made prominent. Under other circumstances, when the growth is undergoing a myxomatous degeneration a section of the growth will present a honey-combed surface, and the hollows of the fibrous structure will be filled with a gelatinous mucoid fluid.

Symptoms.—These benign solid tumours of the breast are usually discovered by accident, the patient's

H—25

attention being rarely drawn to the part by anything
that could be called pain ; an uneasy sensation in the
breast is the most that is experienced. Such a
tumour feels, when discovered, as a hard or fleshy,
smooth or lobulated nodule, upon, behind, **or within**
the substance of the breast gland ; it is generally
very movable, **and** feels like a loose body in the
breast, which may at times be separated from **it ;**
but **should** the growth occupy the centre of the
gland or one of its lobes, this **separation** will not
be possible. As **it** grows, it troubles mainly from
mechanical causes, for it maintains its original features throughout ; it may develop fresh lobules, or
become more rounded in its outline, but **it** will never
do more to surrounding structures than stretch them,
or press them aside. The mammary gland itself, when
these growths are large, may become atrophied, or
pressed away up to one or other **side of** the **new**
growth, according to the lobule of **the breast in**
which the growth originally appeared. When the
tumour is **central or** approaches that position, the
breast will be flattened out over its surface. The
nipple is rarely much influenced by the growth,
and when so it is from mechanical causes. Thus
a tumour occupying the centre of a lobe covered
with breast structure will stretch the ducts, draw
upon the nipple, **and** thus cause either an obliquity **of**
its position, **or** some retraction (Fig. 4, page 177).
Should the tumour be a central one, and press equally
in all directions, the nipple will become either flat
or retracted ; I have seen several examples of this
affection in which this latter condition was present.

Pain.—Very little pain attends **the** growth of
these tumours ; indeed, in the majority **of** cases the
only sources of bodily discomfort **of** which patients
complain are due to the mechanical effects of the
tumour. The chief distress **is mental, viz. the fear**

of what seems so simple becoming eventually a
"cancer," or a source of unknown and severe trouble
unless submitted to operation. In exceptional cases
pain is, however, experienced. Mr. Birkett mentions the fact that in one case intense suffering
was **due** to the filament of **a** nerve being included
in the new growth ; and I **have** certainly found **the**
breasts of some young women **who** have these growths
as sensitive as the simple **and** uncomplicated irritable
mamma.

The *integument* covering the growth **is** very rarely
implicated, since it is elastic, and yields kindly under
a· distension which is slow and steady. Should the
growth, however, be rapid, the skin may inflame, and
even ulcerate, so as to allow the escape of the tumour's
contents. With solid growths such **as we are now**
considering, this result is very rare ; **I have** but **once**
known inflammation to **occur** from distension **in such**
a case. In cystic disease the complication **is** more
common. These tumours never become fixed to the
deeper tissues, and only affect them mechanically by
pressure, although, as I have said, they may be found
to be adherent **to** adjacent portions of the breast.
They do not **as a** rule give **rise to any** lymphatic
enlargement, and should they do so, it will probably
be from some degenerative **or** necrotic change which
is taking place in the growth, or from some irri-
tating effect the growth is producing on **the skin**
or parts around. When lymphatic enlargement
co-exists with a tumour which is believed to be
benign, **the** suspicion of its being otherwise should
be raised.

Rapidity of growth.—The **more** solid the
tumour, the slower **its** growth ; the more succulent and
loose the structure, the more rapid **its** increase. When
associated with cysts, a tumour may enlarge rapidly,
and this enlargement may be the product of increased

effusion, or due to the occurrence of bleeding into the cyst. I have recorded a marked example of the latter cause in the twelfth volume of the Pathological Society's Transactions. On the other hand, these tumours on rare occasions cease to grow or even diminish in size; when it is probable that the diminution in size is due to the absorption of some of the fluid that bathed the free portion of a pedunculated growth.

I have been quite unable to determine the average rate of increase of these tumours; it is so variable. The fibromata and more solid adeno-fibromata rarely attain any size; one the size of a large walnut may be the product of three or four years' growth, and another three inches across represent the growth of fifteen or more years. The more succulent varieties of the adeno-fibromata, although they are usually larger than the more solid, attain a great size in only rare examples. I have removed one of the compact form from a married woman, aged 30, four inches across, that was said to have grown in as many months. In exceptional cases both the solid and succulent varieties may grow rapidly. The cystic forms, as will be shown later on, may, however, increase up to the weight of many pounds. A tumour that grows rapidly with the clinical features of an adeno-fibroma, and is not cystic, is probably sarcomatous.

Multiplicity of and recurrent growths.— It is by no means rare for the breast of a woman to hold two or more of these adeno-fibromatous tumours, and both breasts at times may be equally affected. When more than one exists in the same breast, the tumours are as a rule of the same form and density, but I have on more than one occasion removed from a breast a compact tumour, which would be pathologically described as a fibroma, and at the same time a second tumour of a looser and more succulent kind, made up of glandular elements, to which the term adenoma would be applicable : the former example containing but little, and the latter much, of the glandular structure. In the same way, recurrent growths may vary in individual character, though agreeing in kind; a primary growth **may** thus be of the compact **kind, and** the recurrent of the **succulent ;** in Prep. 2299[45] **of** the Guy's Hospital Museum there are three specimens demonstrating this, the third partaking of the characters of the first.

What has been described as recurrence in any given case may be, in truth, simply the continued growth of a tumour which was overlooked during the first operation on account of its smallness. I have twice seen a second small fibro-adenoma come into **view in the** incision for the removal of the first **growth,** which, had it been undiscovered, would have **grown,** and been called, though wrongly, a recurrent tumour. Tumours that re-appear a few months after the removal of the first growth should be regarded as growths **that had** not been discovered. (*See* pages 108, 109.)

These "*adeno-fibromata*" are found as a rule in **the** young and unmarried, and in the apparently healthy and robust ; they occur, however, occasionally in the aged. The best example of the kind I have seen **was** in a single lady, aged 71, and it had been of twelve years' growth. When I saw her it was ulcerating. I consequently removed the whole tumour and gland, **with an excellent** result. The **specimen** contains **samples** of every variety of this class of tumours ; **some** masses representing well the fibroma, others the most typical and marked **adenoma,** both of the compact and loose forms. The adeno-fibromata are seen in every stage. The more compact form of adeno-fibroma, or fibroma, appears generally in girls about the age of puberty, the gland structure at this age having but

just advanced sufficiently far so as to be ready under
the natural stimulant of conception for a more complete development.   In such subjects, when the
breast is the seat of a new growth, the adenomatous
element is slight, but the fibrous is in excess.   In
maturer women, and specially in those who have
borne children or nursed (when the procreative organs
are active, and the mammary glands are in a state of
"developmental perfection") the most perfect example of adenomatous tumour is to be expected,
since in these subjects the glands have reached their
highest physiological function, and consequently their
highest development, and any new growths originating in connective tissue at this period would naturally
include the active gland structure, and present a more
or less succulent and well-marked adenoma or adeno-
fibroma, the gland elements being in excess over the
fibrous.   They are occasionally met with in men in
the proportion of 1 to 277 cases of adeno-fibromata in
women, as shown by Mr. W. R. Williams, in an able
analysis of 11,100 cases of neoplasms admitted into
four large metropolitan hospitals, during periods
varying from ten to seventeen years, and ending 1883.
In women the majority of these cases begin to grow,
or are first discovered, between the ages of 21 and 30,
although, as shown by my table, almost as many
begin in younger people, at or after puberty ; at
later periods of life they less frequently originate,
although they may be frequently found to exist in
them.

From my notes of 100 cases consecutively observed,
seen, and analysed to make out these points,

> 27 cases were first discovered between puberty and the age
> of 20, that is, during the developmental stage of the
> breast's life.
> 35 cases appeared between 21 and 30 years of age, or
> during the period of its functional perfection.

22 cases appeared between 31 and 40, during the period of
    its maturity.
13 cases appeared during 41 and 50 ; and
3 cases appeared in women over 50, or during the period
    of its functional decline.

46 of these cases occurred in single women ;
39 in the married and prolific ;
15 in the married and sterile.

Single women during their prime are consequently
most prone to be attacked by these growths, and the
married and prolific are the next ; the old maid and
the married and sterile woman being comparatively
free from these benign tumours.

Cause.—No definite cause can usually be ascribed
for the appearance of these growths, although in
quite exceptional cases injury has been assigned. I
cannot say, however, that I have ever been satisfied
with any such explanation, for the tumour has usually
been discovered too soon after the supposed cause to
render the suggestion probable. As a rule, indeed,
these tumours are discovered by chance, some accident
or blow having drawn the patient's attention to the
part, and thus led to the disclosure of the tumour.

Diagnosis.—This should not be a difficult task
in the majority of cases. A fleshy or firm movable
lobulated tumour in the breast of a healthy looking
single or married woman under thirty ; with a mammary gland, its nipple and skin covering, in other
respects apparently healthy ; and with a history of the
slow **and** probably painless increase of the new
growth, or if painful, only so at intervals, or **at**
the times of the catamenial flow, is probably an
adeno-fibroma. The more movable it is in the
gland, the greater the probability of this diagnosis
being correct ; the more fixed it is, the greater
the chance of the tumour being sarcomatous Any
growth that infiltrates the gland, wholly or in part,

must be inflammatory or cancerous. The absence
of fluctuation in any part of the tumour, and the
absence of a marked lobulated or botryoidal outline, excludes polycystic disease; although in every
case of growth believed to be benign found in the
breast of a woman over thirty, the suspicion of its
being cystic should be entertained. The solid or
non-cystic fibro-adenomata grow, as already described,
in capsules, and simply expand the tissues in which
they are placed; they do not spread, as the cancers
do, by either local, lymphatic, or vascular infection,
but cause trouble simply mechanically by their presence; they do not give rise to discharge from the
nipple, as do many of the cystic tumours, and they are
generally found in women who are apparently healthy.

Prognosis.—This is, as a rule, favourable, for
when one of these tumours has been removed there is
every reason to believe that no return will take place.
When a second growth appears soon after the removal of the first, there is good reason to believe that
it is but the continued growth of a neoplasm which
existed undiscovered at the time of the first operation;
for I have on two occasions at least, when an adeno-
fibroma was being removed from a breast by an incision through the breast gland, which covered the
capsule of the neoplasm, seen exposed a small encapsuled second growth, about the size of a pea,
which, if left at that time, would have grown, and
appeared as a second or return growth. Second or
third growths are, however, met with, but such, as a
rule, originate in other lobes than that primarily involved, and cannot consequently be called recurrent,
for such a term refers to the re-appearance of a neoplasm in the position of some antecedent growth,
which I do not believe takes place in one out of a
hundred cases. Many of these tumours after a time
cease to grow, or do so but very slowly; some without

doubt diminish in size, and when this diminution
takes place, the growth is probably cystic, and the
absorption of the fluid about the growth has caused
its apparent shrinkage. To this point attention will
be drawn later on. (*See* Shrinkage of tumours, page
344.) The more fibrous and glandular, or adenoid, the
growth is, the better the prognosis. The more embryonic and cellular the connective tissue elements
(that is, the more the growth approaches an adeno-
sarcoma and deviates from an adeno-fibroma) the
greater the probability of a return.

Treatment.—As a general rule all these tumours
should be removed by excision, since experience
tells us that they will continue to grow with variable
degrees of rapidity ; that a tumour which has taken
months or years to attain a certain size may often
double itself in as many weeks or months, and that
no local treatment by applications and no medicines
have the slightest influence in retarding their development or helping their disappearance.

It is not, however, necessary to remove every
tumour of this kind as soon as it is discovered, for it
may grow so slowly, and be so little in the way, as to
render its removal a matter of small urgency ; at the
same time, such a tumour should be taken away
from any woman who is **likely to** become pregnant,
or who is past the age of 35 ; since with a pregnancy
the neoplasm is certain to grow, as it, with the
gland **itself,** will have increase of blood sent to **it,**
and likewise from nerve causes will receive an im-
pulse towards growth ; and after the age of 35
adeno-fibromatous growths are more likely to **pass**
on to the adeno-sarcomatous, which are far more
dangerous.

The operation itself is neither a difficult nor
dangerous one. It usually consists of a clean incision
through the overlying tissues into the capsule and

then the subsequent enucleation of the growth. **The tumour should be well pressed forwards by the left hand of the surgeon** grasping the tissues deeply on each side of the tumour so as to make it prominent **and** the tissues over it tense; the incision into the capsule of the growth should always be free, and made in a line radiating from the nipple, **as** by this incision there is less harm done to the breast structure **which** surrounds the growth, and the risks of injuring the milk ducts are much diminished. When **the** capsule has been divided, the growth will probably become partly extruded, and its complete **extrusion** will be facilitated by **the** division of any fascia which appears to hold **it** back, and by introducing the handle **of the** knife between the capsule and the growth so as to lever it out of its bed. The attachments of the neoplasm to its capsule are at times very slight, and under these circumstances it is readily enucleated ; when the attachments are **stronger** or broader, they may want division with the **knife,** and under these conditions some vessels may require to be twisted or otherwise secured.

In exceptional cases the neoplasm is so closely connected with the breast tissue as to necessitate the **removal of** the lobe of the gland in which **it is placed.**

All bleeding should be thoroughly arrested before **the wound is** closed, by the torsion of every vessel of any size, and the temporary introduction into the capsule or cavity of the wound of a sponge wrung out of hot iodine water. Its edges should then be carefully adjusted by means of fine sutures and strapping, **and** the wound dressed, a provision being carefully made **for its** drainage for twenty-four hours. A little gentle pressure, applied either through iodoform gauze, salicylate wool, or a soft sponge rendered aseptic **with iodoform or** boracic acid, and bound on with a bandage, is most useful in checking capillary oozing and venous hæmorrhage.

In the majority of these cases the wound heals by quick or primary union; where this wished-for result is not secured, the cavity must be daily washed out with iodine water, or some other antiseptic wash, well drained, and carefully dressed with such antiseptic dressings as are favoured by the surgeon.

Breasts, after having been subjected to an operation for the removal of one of these benign neoplasms are rarely so injured as to prevent their performing their physiological functions when occasion calls. In the larger forms of tumour and of operation this result may ensue; but when the neoplasm is of moderate dimensions it is not to be anticipated.

Should the tumour have been allowed to attain a large size, it may be impossible to separate breast gland from growth; under such circumstances it may be necessary and expedient to excise both. It must, however, be stated that large tumours, even up to seven pounds, may be removed from a gland, and the latter left intact and capable of performing its true functions; and under these circumstances it is the surgeon's duty to attempt to save the gland, even when it may appear hopeless. These principles of practice are clearly applicable to cases met with during the child-bearing period of life. At later periods the object for retaining the breast does not exist, consequently the practice based upon it is not necessary.

To render the scar of the operation as invisible as possible, Gaillard Thomas, of New York,* when the tumour is of moderate dimensions, makes his incision in the fold which unites the lower hemisphere of the breast to the thorax, and having dissected the gland from its deep attachments, removes the tumour by an incision made in the under surface of the breast. The

* *New York Med. Journ.*, p. 337; April, 1882.

wound is then carefully drained and **treated, and the** subsequent scar is **very** limited.

It occasionally happens that a breast which has been the seat of an adeno-fibroma becomes, **in** later years, the subject of a cancer. The secondary neoplasm must be regarded as a new growth altogether, and as having only a very slight connection with the first, that connection being probably the same as exists between a cancer and a scar in other parts.

To more fully illustrate many of the practical points mentioned in these pages, the following brief **notes of** cases, extracted from my note-book, will be read with interest.

The *first* case refers to the fact that in making a section of breast tissue to enucleate an adeno-fibromatous growth, a second smaller tumour was exposed.

1. *Adeno-fibroma of breast; excision; second tumour discovered by the incision made to remove the first; removal of second; recovery.*—Harriet P., aged 30, single, a servant, came under my care on September 23, 1874, for tumour in her left breast. It had appeared nine **months** previously as a small movable lump, which **steadily** increased. At present it measures three inches by two, is situated below the nipple, is movable, nodulated, and firm. On the 25th an incision was made over the tumour, in a **line** radiating from the nipple, **into its** capsule, **and** the growth enucleated. When **this** was done, a second tumour was discovered placed deeply in the gland beneath the one that had been removed. This was then enucleated, and a good recovery followed. A precisely similar case **to this** came under my care in the person of **Sarah V.,** **aged** 25, in March, 1882.

The *second* case illustrates a **very important point** in the life history of an adeno-fibromatous **or** other tumour; and that is, the enlargement of lymphatic glands, **and** their **subsequent** subsidence after the removal of the primary growth.   The case is likewise an example of multiple tumours.

2. *Two adeno-fibromata in one breast, with enlarged lymphatic gland ; subsidence of latter after removal of tumour.*—Sophy S., aged 30, a childless married woman, came to me on May 19, 1850, with two hard lobulated tumours on the axillary half of the left breast, which had been growing for three years.   One was as large as a walnut, the second as a nut.   Both moved freely in the breast.   There was also an enlarged lymphatic gland in the axilla.

May 28. The tumours were enucleated through one incision, and were excellent specimens of the adeno-fibromata ; the gland elements predominating. The patient made a good recovery after the operation. The enlarged lymphatic gland disappeared. **This** patient was quite well four years later.

The *third* case is an example of double tumour in a breast, the two tumours being good examples of two of the varieties of adenoma.   It shows, moreover, the effects of pregnancy upon the growths, and the value of saving the breast in the operation.

3. *Adenoma of* **breast,** *associated with adenofibroma, complicated with pregnancy ; suckling with the affected gland; operation; recovery.*—Frances F., aged 22, came under my care March 15, 1869, with a large tumour in her left breast of two years' standing, and a smaller one of twelve months'. She was then six months pregnant.   During the pregnancy the tumours grew fast, as they did during lactation, which she carried on with the affected breast, after a natural labour, for three months, when she weaned the infant. On November 7 one tumour was as large as a cocoanut, and the second the size of a small orange.   The latter tumour was much harder than the former. On January 3, 1870, I removed the tumours and left the breast intact.

The larger tumour, as seen in Fig. 2, page 92, and Plate I. Fig. 3, was a fine specimen of the looser kind of adenoma. The smaller was more of an adeno-fibroma; the fibrous tissue predominating over the glandular. This patient subsequently suckled with the breast, and was in good health five years after the operation. This patient's sister came to me in 1870 with well-marked so-called hypertrophy of her left breast. The gland was twice the size of the right. She was single, and 21 years of age.

The *fourth* case illustrates the effects of pregnancy and lactation in the growth of an adenoma.

4. *Adenoma of eighteen years' growth, with rapid increase during lactation.*—Mrs. H., aged 38, a widow, the mother of one child three years old, came under my care on June 12, 1867, with a large nodular and fairly elastic tumour, the size of a cocoanut, in her left breast. The breast gland could be separated from the tumour without difficulty, and the nipple was natural. The tumour had been growing for eighteen years, slowly for fifteen, but rapidly after her last pregnancy and lactation. She would not consent to have it removed.

The *fifth* case is full of interest, since it is an example of adeno-fibroma in a woman, aged 62, and was associated with a retracted nipple, a sign which is too readily accepted as an indication of carcinoma. The new growth also was said to have followed an injury.

The *sixth* case likewise illustrates one of the same points, since the tumour was associated with a retracted nipple. In it, as in the former case, the growth was placed in the centre of the breast gland.

5. *Adeno-fibroma of the breast in a woman, aged 62, following a blow, associated with retracted nipple.*—Dorothy S., aged 62, came under my care on July 23, 1884, with a hard lobulated tumour occupying the

centre of the left breast, associated with a retracted
nipple, but no axillary glandular enlargement. The
tumour had been growing for eight months, and had
followed a blow received upon the part one month
previously.

July 30. **The** tumour was removed by making a
free **incision into its** capsule, **in a line** radiating from
the nipple. It was encapsuled and turned out with
the greatest ease. A good recovery ensued. **The
tumour was** an excellent example of adeno-fibroma.

6. *Adeno-fibroma of the breast, with retracted nipple,
in the daughter of a woman who had carcinoma of the
breast.*—In 1878 I removed from the left breast of
Miss N., a healthy girl, aged 18, an adeno-fibroma of
six months' growth, which was placed in the centre of
the breast, and gave rise **to** retraction of the nipple.
The case did well. I had two years previously **re-**
moved the right breast of her mother, who was then
well, for a rapidly growing carcinoma. Her mother's
sister had likewise had cancer of the breast.

The *seventh* case is quoted as a rare example of
keloid attacking the scar of an old breast operation,
and the still rarer absence of return after the excision
of the keloid growth.

7. *Adeno-fibroma of breast; excision;* **eight years**
later keloid of scar; **excision** *; no return five years
later.*—Miss M., aged 34, a healthy lady, consulted me
on December 30, 1870, for a typical adeno-fibroma of
her left breast, which had been growing for six months.
It was hard, bossy, and movable. It measured four
inches **across.** I removed the growth without injuring the breast, and the case did well. **On**
February 24, 1879, that is, over eight years after **the**
operation, this patient came to me with a genuine
keloid tumour in the scar of the old operation, which
had been coming for seven months. Dr. Sidney Turner
of Sydenham, under whose care she was, excised the

keloid with a good margin of skin, and on September 25, 1886, he reported the lady as being well.

This case is doubtless interesting, first from the fact that a keloid growth should attack the scar of an operation eight years after its performance; and second, that on the removal of the keloid there should be no return.

The *eighth* case is, in every way, an admirable example of adeno-fibroma of the breast. It began early in life, originated during pregnancy, grew rapidly during several puerperal periods, and shrank after lactation to its former dimensions. When it was excised it weighed seven pounds, and in this operation the breast was uninjured, as it was able later on to perform its physiological functions. The clinical history of the case extended over ten years.

8. *Case of adeno-fibroma of seven pounds weight regarded as one of hypertrophy, and removed; the breast was saved, and secreted freely in later years.*— Eliza K., a healthy looking married woman, aged 24, was admitted into Guy's Hospital, under the care of Dr. Ashwell, on October 15, 1840, with a large tumour apparently involving the *right* breast.

She married at the age of nineteen, in due course had a child, and nursed it with both breasts, but had to wean the infant on account of sore nipples. The right breast, as a consequence, became the seat of an abscess, which opened in many directions, and then healed. Within eight months of her first confinement she again became pregnant, and in about the twelfth week she perceived a lump the size of a hen's egg in the axillary border of her right breast; it was neither tender nor painful, but it steadily grew, so that on the third day after a natural labour the tumour measured eighteen inches in circumference. Mrs. K. nursed only with the left breast, and the right began to decrease in size, so that in the eighth month of

lactation the growth was not larger than an orange.
At this period she became pregnant for the third
time, and at the end of the third month, as in the
second pregnancy, the tumour began again to enlarge,
but more rapidly, though without pain ; the tumour

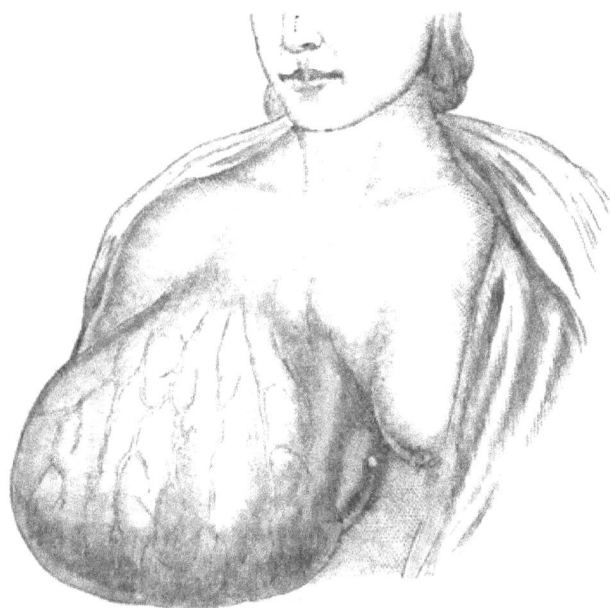

Fig. 3.—Adeno-fibroma of Breast.

then measured twenty-three inches in circumference.
At this time she came into the hands of Sir A. Cooper,
who punctured the tumour and evacuated two ounces
of curded milk, the wound soon healing.   She became
pregnant again, was prematurely confined, and did not
suckle, as the child died twenty hours after birth.
The tumour, however, on this, as on the former
occasions, became smaller, but not to a great extent ;
it dwindled down, however, from twenty-three to
fourteen inches.   A fourth pregnancy then occurred,

I—25

when, as before, the tumour increased, and soon attained a measurement of twenty-nine inches, and a weight of nearly twenty pounds. At this time she came under the care of Dr. Ashwell. A natural labour soon followed, and four days after parturition, **Dr. Oldham, who was watching the case, reported that the** circumference **of the tumour had increased two inches and a** half, **so** that it measured **thirty and a half inches.** Four weeks after labour the tumour had **decreased five** inches; she did **not** suckle with the breast, although she did with the **other**; on applying pressure to the breast, a little milk exuded from the nipple. The engraving above represents the appearance of the growth, which was lobulated, soft, and **elastic.** It measured twenty-nine inches round, and **weighed** nearly twenty pounds. Dr. Ashwell regarded **the** disease as one of hypertrophy. In 1843 the **patient** passed into the hands of Mr. Stanley, who removed the tumour on July 16, 1843, leaving the normal **breast, seen** in the figure, on the sternal side of the **tumour,** unmolested. The tumour then weighed only **seven pounds.*** A further note of this interesting case is found in the Guy's Hospital Reports† by **Mr. Birkett, who writes :** " I have seen this patient **several times** since the operation. She has **given birth to three children, and suckled them with the breast from which the tumour was removed.** When I last saw her, which was seven years after the operation, she was in good health, and quite free from any disease. I have, through the kindness of Mr. Stanley and with the assistance of the patient herself, identified the tumour in the museum of the college **(Prep.** 399) with the woman whose history is recorded by **Dr.** Ashwell, as given above.

---

* *See* Prep. 399, Museum Royal College of Surgeons. Drawing : Guy's Museum, 401^{50}. Cast 2789^{59}.

† Guy's **Hosp.** Reports, vol. i. p. 144; third series; 1855.

PLATE II.

Fig. 1.

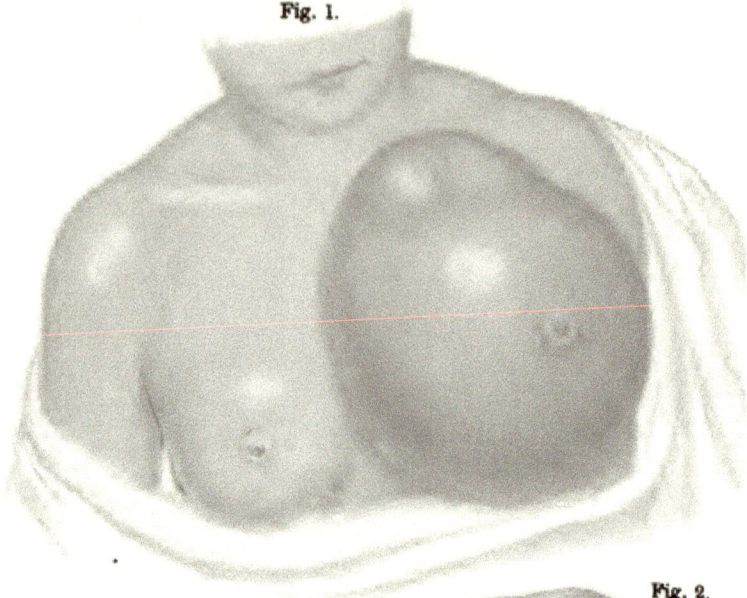

Fig. 2.

Fig. 3.

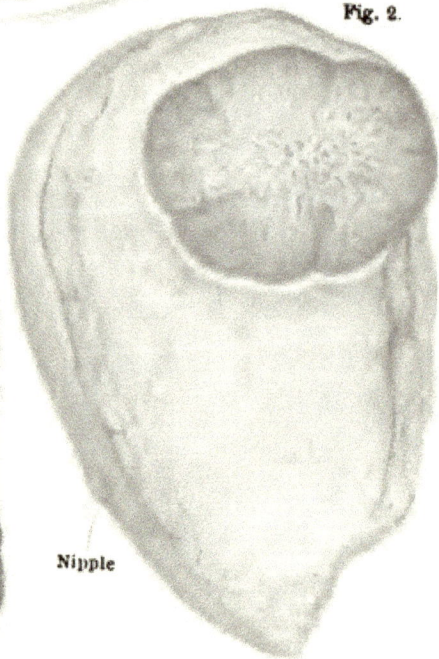

Nipple

Nipple

SARCOMA OF THE BREAST.

1. Sarcoma.   2. Section of harder Variety.
3. Section of softer Variety.

CHAPTER XI

ADENO-SARCOMATA.

The sarcomata of the breast gland: an encapsuled growth.—Sarcomatous tumours are new growths that originate in connective tissue structures, and are composed of embryonic connective **tissue** elements. The cells are, as a rule, irregularly arranged in an intercellular substance, through which the blood-vessels permeate.

They are probably formed from the connective tissue cells, and the cells assume sometimes **a** spindle, sometimes a round, and but rarely a giant shape ; the name of the growth being determined by **the** marked predominance of any one kind of cell elements.

These growths, **in** only exceptional **instances, infil-**trate a **part,** as do the epithelial or carcinomatous, but are more or less *encapsuled*, the capsule varying considerably in its density, this density being apparently much determined by the rapidity of the tumour's growth (Plate II. Figs. 2 and 3).

The tumours, moreover, vary much in their *consistency ;* some **are** firm and approach the fibroma in feel and appearance ; in such, the fibrous tissue is abundant, and the cell element comparatively small (Plate II. Fig. 2). The cells, moreover, in these firm growths will probably be of the spindle shape, the cells of the **connective** tissue growth having developed into fibres, **which form** bundles, fasciculate, and **con-**tain within their **meshes** undeveloped spindle cells. These growths form **the** "*spindle-celled*" group of sarcoma, which have been described by the older writers as the "fibro-plastic," or recurrent fibroid, and so described because they are prone **to** recur, after removal, in the part from which they originally grew, or in the neighbouring parts.

These spindle-celled sarcomatous growths on section usually present a white, glistening, semitransparent surface, which bulges to form a convex outline. They may be homogeneous, or to a degree lobulated, and on scraping the cut surface it will yield the oat-shaped cell, with an oval nucleus.

The soft or more succulent varieties of sarcoma are probably the *round-celled* kind, which seem to be composed chiefly of cell elements, held together with but little fibre tissue. These are well supplied with blood, and in some cases are so vascular that they pulsate, and become, from slight injury or other cause, the seat of blood extravasation. Such tumours are of rapid growth, and present in section a white medullary or a more or less blood-stained convex surface (Plate II. Fig. 3). In a clinical sense there are, however, great difficulties in correctly diagnosing these forms of tumour.

In some, again, where much blood has been extravasated, the cell elements are difficult to find; indeed, the tumour seems to be but a blood clot. From such tumours of the breast I have known a quart of fluid, coffee-ground, or almost pure blood, drawn off.

Many round-celled sarcomata have spindle cells and fibres mixed with them. Many lie simply embedded in the tissues, and turn out like loose bodies; such have a very glistening convex surface in section, and seem to a degree translucent. They are at times, though rarely, multiple. All forms of sarcomata, besides recurring locally, tend, as is well known, to disseminate themselves by the blood-vessels. At times even they may spread through the lymphatics; thus local recurrence, vascular and lymphatic infection, conspire to produce the most inveterate form of malignancy.

When any of these connective tissue tumours originate in the breast, which is composed of gland structure, in addition to connective tissue, some of the glandular elements, either tubes or their cæcal terminations, are mixed up with them.

Under these circumstances, the anatomical structure of the growth becomes modified, and thus it is that the more expressive term "adeno-sarcoma" seems to be applicable. In some cases these glandular elements will form but a very small part of the tumour, and will be detected scattered about its tissue only after a very careful examination, whilst in other cases they will be met with in abundance, mixed up with the other elements. It is this mixture of gland structure with the sarcomatous elements that specialises the connective tissue tumours of the breast, and more than justifies the retention of the term "adeno-sarcoma."

It is difficult, if not impossible, in the present state of our knowledge, to say in what proportions of sarcomatous tumours the round-celled, spindle-celled, or giant-celled neoplasms predominate. My own material, although extensive, is not conclusive upon the subject; indeed, I should say, after close examination of notes of cases, that the majority of them are of the mixed form, that is, made up of round and spindle cells in different proportions, the giant-celled variety being rare, and the melanotic very so.

In a large proportion of cases, I should say in half, the adeno-sarcomatous tumours of the breast are associated with cysts, when they are known as cystic sarcomatous tumours. To these, special attention will be directed in chapter xvi. In the present chapter attention is alone drawn to the more solid varieties.

Gross has attempted to strike an average as to the relative frequency of the different varieties of solid sarcomata, and states that sixty-eight per cent. of one hundred and fifty-six cases collected and seen by him, were of the spindle-celled variety, twenty-seven per cent. of the round-celled, and five per cent. of the giant-celled. By this estimate, two out of three cases **are** spindle-celled tumours.

From a clinical point of view there is, however, no great necessity for separating the different varieties of tumours, beyond recognising the general fact that the more elementary the cell structure of which it is composed, and the greater the proportion of cell elements which build it up, the more rapid will be its growth, and the softer its structure ; the greater also will be the probability of a speedy return of the tumour after removal, and the stronger the fear that secondary growths will take place in the viscera or other parts. On the other hand, when the cells tend towards spindle shape and are not numerous, when **the** tumour is generally **firm in consistence and slow in** its growth it is less likely to return rapidly after removal, as well as to affect other parts. In short, **whereas** the former or softer type of tumour approaches the carcinomata in its natural course, the latter or firmer type tends to follow the course of the benign or non-infective varieties of tumours.

The sarcomata in the breast differ also in themselves as much as the sarcomata of connective tissue are known to do ; and they may appear either as firm or soft succulent growths, with little or much blood in them, or almost as blood tumours.

The amount of cell elements, as well as the character of the cells, will probably vary in every example. Where the cells are few and connective tissue fibres abundant, the structure of the growth will be firm ; where the reverse holds good, and the cell **elements** predominate, it will be soft and medullary.

In the matter of vascularity, these embryonic connective tissue tumours vary materially ; some have a perfectly white and creamy appearance, whilst others possess a pinkish, semitransparent surface. A few are very vascular, and show on section, scattered through their tissues, innumerable vessels ; whilst in some cases the growth seems to be so full of blood as to appear as a blood cyst, or tumour ; such cases as **these** have suggested to pathologists the probability **that** the embryonic connective tissue cells themselves possess the power of forming blood-vessels and blood corpuscles, as a pathological process, in the same way as they are known to do in fœtal life as a physiological process in the embryonic mesoblast; the blood-forming function of the connective tissues of the mesoblast in embryonic life having been re-awakened, after its normal period of activity has long past, to form **a** sarcomatous tumour of a sanguineous type. Dr. **C.** Creighton has cleverly worked out this view.

These sarcomatous neoplasms **in their** clinical features simulate the more benign "adeno-fibromata."

Like the adeno-fibromata, they are encapsuled, but they are far more intimately connected with the **breast** than are those growths ; they may be said, indeed, to be more closely connected with their capsule to the breast than are the adeno-fibromata, and they can rarely be enucleated from their beds in the same way.

As primary growths they never infiltrate the breast, in the characteristic way of the carcinomata. **As** secondary or recurrent tumours they do in exceptional **instances.**

They grow and separate tissues like the adeno-fibromata. They also stretch the integuments covering them in, and even cause their rupture ; and when this event has taken place, the neoplasm bursts forth in **all** its force, and shows itself as a soft, sprouting, fungating, and bleeding mass, which the old authors called a fungus hæmatodes.

These neoplasms grow rapidly, far more so than the adeno-fibromata or the majority of cancers ; a tumour of this kind may attain the size of a cocoanut in three or less months. At times they grow slowly at first, and then suddenly increase at a rapid rate. They also speedily undergo degenerative changes and break down, showing then in section great cavities ; at times they even slough. The outline of these tumours is as a rule more uniform than an adeno-fibroma, it is rarely so lobulated ; it usually originates as an isolated growth; it is rarely, if ever, multiple. It is never so movable within the gland as is an adeno-fibroma, but seems part and parcel of the lobe of the gland in which it originated. It is usually as movable with the gland as an adeno-fibroma. It is rarely associated with any lymphatic glandular enlargement, although in exceptional cases it may be ; this complication is more likely to occur when the skin becomes much distended, and consequently irritated by the growth, than under other circumstances. In some of the worst examples of this disease the lymphatic glands are never involved.

"Sarcoma," writes Gross (p. 22), "however, may extend along the blood-vessels and invade the adjoining tissues without its capsule being necessarily destroyed. Hence, during their further growth and extension, sarcoma, like carcinoma, exhibits malignant attributes, as evinced in the latter by the continuous growth of the cells into the coverings of the mammæ and the subjacent structures, and by their transportation to associated lymphatic glands and the viscera, where they proliferate and supplant the natural tissues ; and in the former by the same phenomena, with the exception of the conversion of the lymphatic glands into secondary growths."

These sarcomatous growths likewise simulate the carcinomatous in their tendency to recur, and, what is more, in their rapid tendency to recur. It often happens that a second tumour will appear before the wound of the operation for the removal of the first has closed ; so rapidly, indeed, may this apparent recurrence ensue, that the surgeon is led to think that what is regarded as a recurrence may have been but **a** continued growth of some part of the neoplasm **that** had been left behind, or some tumour which had not been discovered. In many cases he will doubtless be right in his suspicion ; but in others all evidence upon the point is deficient, and the reappearance of the growth must be regarded as a recurrent growth. These local recurrences may continue for many times. I have in one case already operated seventeen times in the course of three years, and the disease is yet quite local. (*See* case 1, chapter xx. page 335.)

In some cases the disease not only returns locally, but like the carcinomata involves other organs, and particularly the viscera. Gross states "that mammary sarcoma recurs locally in 61·53 per cent. of all instances, and that it gives rise to secondary deposits in distant organs in 57·14 per cent. ; " whilst "in carcinoma 88·35 per cent. of the local recurrences are met **with in** the first year, while only 50 per cent. of the cancers affect distant parts."

It is, however, to be remembered that Gross draws his experience from selected cases, in which the proof of the sarcomatous return of the growth was determined by histological examination.

These sarcomatous growths rarely attack the breasts of young women, but rather of those **over** 30 or 35, and they are not uncommon **in** women about fifty years of age. They are found also in male subjects ; out of 68 examples of sarcomata of the breast consecutively observed in four large London hospitals, according to Dr. W. Williams, two were in men.

Diagnosis.—The diagnosis of an adeno-sarcoma

from an adenoma or adeno-fibroma is difficult, and in many cases clinically impossible. Its true nature is to be determined better by its progress and clinical history than by its physical features. A lobulated, slow-growing, firm, fleshy, movable growth in the breast of a young woman is probably a "fibroma," and not a sarcoma ; whilst an ovoid, smooth, elastic tumour, of somewhat rapid growth, in the breast of a woman over thirty, is probably a sarcoma. The younger the patient, the more lobulated the neoplasm, and the slower its growth, the greater the probability of the tumour being a fibroma. The older the patient, the smoother the outline of the tumour, the more elastic its feel, and the more rapid its growth, the greater the probability of the neoplasm being a sarcoma. When the integument covering in the tumour, which was solid, is ulcerated, and the growth appears through the orifice as a sprouting, fungating mass, the disease is probably sarcomatous, although, should the disease have been cystic, this conclusion would not be so clear. As to the relative frequency of adeno-sarcomatous and adeno-fibromatous tumours, out of my last hundred cases of these collective growths but four were pronounced to be sarcomatous.

"A tumour," writes Gross (p. 99), " of soft, elastic, apparently fluctuating consistence, which attains the volume of an adult head in a few months, can scarcely be anything else than a small-celled sarcoma. On the whole, the diagnosis is based upon their indolent origin, lobulated outline, rapid increase, large dimensions for the period of their existence, freedom from lymphatic involvements, and marked tendency to ulcerate; upon the not infrequent discoloration of skin, enlargement of the subcutaneous veins, and possibly, elevation of temperature ; upon the suffering which they awaken late in the disease, and upon their greatest frequency after the thirty-fifth year."

Treatment.—These neoplasms cannot be taken away too early; they should be excised **as** soon as they are discovered. They should not only be excised, but the parts around them should likewise be taken away, that is, the lobe in which the growth **originated.** The whole gland should only be removed **when it** seems to be involved in the disease; it is better, however, to err in these cases by taking away **too much** than too little.

Recurrent growths should likewise be dealt with in the same thorough way. These are at times multiple, and when so, the operation must be extensive. I have removed recurrent growths from one patient seventeen times in three years, and at the last operation, in 1887, the disease was still local. Indeed, there is good reason to believe that by removing the local affection, even after many recurrences, the **disease** is prevented from becoming a general one, **and in**volving internal organs. In a case of small spindle-celled sarcoma, Gross succeeded, after removing fifty-two tumours by twenty-three distinct operations (the **last few** of which included portions of the pectoral **and intercostal** muscles, in a period of four years and a half) in checking the reproductions, and the patient was perfectly well nearly eleven years subsequently. Gay, Birkett, Heath, S. D. Gross and Haward have likewise reported cases, showing the advantages of repeated operations in cases of recurrent sarcoma.

In many of these cases, after removal of the tumour, it is wise to leave the wound open, and let it granulate, after having well swabbed its surface with a twenty per cent. solution of chloride of zinc. I have thought that this plan of treatment destroyed such cell elements as might have remained after the operation, and in this way retarded, if not prevented, recurrence. When lymphatic glands are enlarged they should be taken away, but the prospects of anything like a cure under these circumstances are very feeble. Indeed, when the lymphatic glands are involved in sarcomatous growths, the prognosis is as bad as possible, and the surgeon should consider carefully as to the expediency of advising any operation. It may safely be said that it can be in only exceptional cases a justifiable measure.

By way of further illustration of the subject, the following cases may be quoted :

1. *Adeno-sarcoma of breast ; excision of tumour ; recurrence of growth; ability to suckle with breast between while ; excision with the breast ; well two years later, five years after the appearance of growth.*

Elizabeth M., aged 36, was admitted into Guy's Hospital under my care on the 14th of July, 1880, and was discharged on the 16th of August, 1880.

The patient is a married woman, and has eight children, seven of whom are living. She was in Lydia ward three years ago under my care, when I removed a tumour from her breast.

Her father and mother are both alive ; there is no history of cancer with them ; one of the patient's brothers died of tumour of the liver.

The old report of her case describes her as a healthy looking, well-nourished woman; she entered the hospital 7th June, 1877, and then had a large tumour occupying the left breast, but extending chiefly downwards, and outward towards the left axilla. There was a scar to the outer side of the nipple, which was caused by a surgeon making an incision some weeks previous. The circumference of the left breast was 18 inches. On June 19th, 1877, patient was put under chloroform, and Mr. Bryant made a transverse incision eight inches long, reaching from the outer boundary of the tumour, inward and towards the level of the nipple, so that the growth was removed without extirpation of the breast. A large axillary gland was also removed; the great size of the arteries was very noticeable, one of these being as large as the radial. The growth weighed 2lbs. 4ozs., and was composed of a number of cysts, some containing a milky fluid, and others more solid contents. The patient left the hospital on 18th July, with the wound almost healed, and in a fairly satisfactory condition.

Since leaving the hospital, the patient has had one more child, and she has been able to use the left breast in nursing; the child was weaned about twelve months since; about this time she first noticed a small lump just above the nipple, which has been increasing ever since.

On admission there is a large lobulated tumour, divided into two chief parts, involving the left breast. One of these parts occupies the central or upper half; the other, which is the smaller of the two, is placed on the axillary side of the gland. The tumour is above the level of the old cicatrix. It is freely movable, with a lobulated surface, elastic to the touch; hard in some places and moderately soft in others, and it is cystic in character; the skin over the tumour is not involved. One boss which is most prominent, and placed just above and to the inner side of the nipple, is inflamed, and painful. On July 20th, 1880, the tumour, with the breast, was removed, and a good recovery followed.

The tumour was everywhere encapsuled, and was composed of many lobules, separated from one another by fibrous septa. Some of these lobules grew into the walled cysts, which contained little or no fluid, their walls being in contact. The growths in these cysts were in some cases lobulated finely on the surface, whilst in a few they resembled in colour and smoothness a mucous nasal polypus, although in consistence they were a little firmer. The majority of the lobules showed on section a finely lobulated appearance, the outlines of the little lobules being crescented, some having a distinct cavity in the centre. No communication could be traced between the various cysts, nor could any be traced directly to the nipple. Some parts of the growth were firm, hard, and fibrous; there were no blood cysts or **any** translucent-looking material, in the growth.

In some of the nodules (felt to be elastic during life) **there were** many elongated slit-like spaces with smooth walls, the bounding material being acinous.

In Sept., 1882, this woman was still well.

2. *Spindle-celled adeno-sarcoma of breast; tumour weighing two pounds on removal; recovery; patient well six years later.*—Matilda H., aged 49, the mother of three children, the youngest being 18, came under my care on Oct. 7th, 1874, with a large **tumour** apparently involving the right breast. The tumour had been growing for two years, and when first discovered was about the size of a nut; it has grown steadily, but for the last two months rapidly. It has never been the seat of much **pain; an** occasional dart of pain through it is all she has felt.

When seen, a smooth globular swelling occupied the position of the right breast. The skin over it was stretched, and the circulation through it was impeded, as shown by the enlarged veins. The nipple was flattened, and the axillary lymphatic glands were natural. On October 13th the tumour and breast were excised, and a good recovery followed. The tumour was found to have been placed behind the breast, or rather in the posterior part of the breast. It was closely connected with the gland, but not infiltrating **it.** On section it was made up of fibro-cellular elements, containing within its meshes spindle cells. It weighed two pounds.

Six years later this patient was quite well.

3. *Adeno-sarcoma (spindle-celled) of breast; excision;*

cure ; patient well two and a half years **later.**— Fanny C., a married charwoman, aged 50, the **mother** of five children, all of whom she suckled, came **under** my care Jan. 25th, 1878, with **a** tumour in the upper part of her right breast, which she had accidentally discovered six weeks previously. It was then the size of a pigeon's egg, smooth and painless. When coming under care the tumour had much increased ; it was about two and a half inches in diameter, firm and fleshy to the feel, and fixed in the axillary lobe of the right mamma. The skin over the tumour was healthy. Nipple natural, axillary glands not to be felt. An adeno-sarcomatous tumour was diagnosed, and removed, with the breast, on Jan. 29th, and the woman did well. The temperature never rose above 99° after the operation. Two and a half years **subse**quently she was known to be well.

The tumour, on examination, was found to be **en**capsuled, but closely connected with the breast. **It** was a semitransparent, succulent, very vascular neoplasm, and under the microscope was clearly a spindlecelled sarcoma.

4. *Adeno-sarcoma of breast ; excision of the gland and tumour ; two years later the opposite breast became involved ; this was excised ; and two years later still the patient was well.*—Mary S., aged 33, **a** single woman of healthy aspect, came under my care in April, 1877, with a swelling in her left breast, which she had detected as a small lump eight months previously. When seen, a tumour was readily made out to exist in the upper and outer border of her left breast, which was so closely connected with the **gland,** that both moved together. The nipple with the skin over the tumour **was** natural, and the lymphatic glands were not enlarged. On May 15th an incision was made into the tumour, and as it was found to be intimately connected with the breast gland, both were

removed. The patient **did** well. **The growth was** found to be a spindle-celled sarcoma.

In August, 1879, that is, two **years after** the operation, the patient re-appeared **with a** growth in the right .breast, which had been increasing for one year, and followed the same course as the one in the left. It was placed in the upper part of the gland, and seemed to be part of it : it measured about two inches across. On August 12th the breast and growth were removed, the patient doing well. The tumour was precisely of the same character as the former one. Two years later the patient was well.

5. *Round-celled adeno-sarcoma of both breasts; removal of both ; patient well two years later*.

Mrs. G., aged 24, the mother of one child, consulted me in December, 1868, for a swelling the size of an egg in her right breast, which she had discovered during her only pregnancy, four months previously. She had been suckling, when I saw her, three months. The swelling was fleshy, smooth, and painless, and was clearly growing fast. Weaning the child was advised. By July, 1869, **the** tumour had increased to nine inches by seven in diameter ; the skin over it was healthy ; nipple natural ; axillary glands uninvolved. It was then removed, with the breast, by the late Mr. Ashforth of Rutland, and sent to me for examination. It was a soft, succulent, pinkish, homogeneous growth which, on section, exuded a glutinous fluid. It was made up of round cells, and some adenoid tissues. The patient did well.

Four months later a like growth appeared on the left, or opposite breast, with some enlargement of the axillary glands of the same side. These I removed, with the breast at once.

One year later I saw this lady with an excellent cicatrix, and after the lapse of another year she was reported as well.

6. *Large cystic sarcoma of the left breast; amputation; recurrence; exhaustion; result, death.*—Emma S., aged 52, a cook, was admitted into No. 3, Lydia ward, under the care of Mr. Bryant, on December 4th, 1883, and discharged on the 15th February, 1884; she was re-admitted on the 2nd April, 1884 and died on the 6th May, 1884.

The patient is a single woman; her father and mother died of old age; she has one brother and five sisters, all healthy. Fifteen years ago patient had small-pox, but since has been quite healthy until ten months ago, when she had bronchitis. About this time she noticed a small hard lump, the size of a kernel, in her left breast, just above the nipple; the growth slowly increased; it was not painful; seven months later it had attained about the size of a fist; no aggravation of symptoms took place during the periods of menstruation.

On admission, the patient is a fat, healthy looking woman; there is a large-sized tumour occupying the position of the left breast which measures 30 inches round its base, $16\frac{1}{2}$ inches across the nipple from above downwards, and $17\frac{1}{2}$ inches from right to left in the same direction (Plate II. Fig. 1). From the measurements it will be noticed that the growth has doubled in size in the last two months. The surface generally of the tumour was congested, with enlarged and engorged subcutaneous veins over the upper two-thirds of its surface; the skin was tense and shiny, and over several areæ the tumour was more prominent and the skin blue, with a mass of dense network of enlarged veins; there were two large bosses above the nipple, which were raised, measuring 3 inches by 3. Areola and nipple were somewhat stretched out; there was no evidence of fluctuation; the surface above the nipple was tender and painful to the touch, and gave evidence of indistinct fluctuation in several spots; there was no involvement of

J—25

the axillary glands; the patient complains of a darting pain at times through **the** tumour; the temperature, just after her admission, was 99°.

On Dec. 7th Mr. Bryant removed the tumour with the breast, and the case ultimately did well; the patient leaving the hospital convalescent.

The patient was re-admitted on April 2nd, **1886.**

History of the secondary growth.—The patient **noticed** round and over the old wound, before it had entirely healed, some return **of** the growth; on re-admission the swelling had extended all over the left **breast, and** at an elevated spot, **near** the axilla, it **had** broken down. The discharge from the growth was very small at first, but it has since gradually increased, and pain has accompanied it more or less all the while. She has some nerve pain down the arm to the elbow.

Condition on re-admission.—The **patient looks** very healthy, but this has been her general appearance all along. The seat of the former tumour is involved with new growth, but there is no lymphatic glandular enlargement. The tumour extends externally to the anterior fold of the axilla; internally as far as the middle of the sternum; above as high as the upper margin of the first intercostal space; and **below** it runs downwards to the eighth **rib.** It presents two sloughing surfaces, **one about two** inches **square, over** the sternum, **at the** upper and inner angle; the growth **being composed** of four nodules which have united and **formed one** mass, the surface of which is sloughing.

About two inches external to this is another **single** nodule, which has a sloughing surface; **the edges of the** two are clean cut and well defined, **and over the** whole circumference there are various nodules, which are soft and rather painful. The skin appears brawny and **of a** red colour. The whole mass moves freely over the ribs and deep parts. The patient's powers steadily failed under the exhausting effects of the discharge from the broken-down growth ; and she died on May 6th. No secondary growths were found.

Melanotic sarcoma of the mammary gland is a very rare affection, and as a primary disease I believe it to be unknown. In the two following cases it was of a secondary character.

7. *Melanotic sarcoma of the breast, **skin**, and axillary **glands**, following the removal of a growth originating in a mole.*—Lydia M., aged 36, a married woman, the mother of one child, nine years old, came **under** my care **in** December, **1857**, with **her** right breast covered and filled with "lumps" the size of **nuts ;** the **skin** tumours were evidently of a melanotic **nature, and** extended **to** the integument covering the sternum and abdomen, some few being in **the** back. The axillary glands were much enlarged. **On** going **into** the history, this woman had had **a** black tumour **the size of a** walnut removed ten months previously from **her** left fore-arm, which had originated in a mole, **and had** been growing one year. The **cicatrix** of the operation wound was sound.

8. *Melanotic sarcoma of breast following the excision of a melanotic tumour originating in a mole over the sternum.*—On October 18, 1867, I was consulted by a **Mrs. M.**, a married childless woman, aged 55, from **whose** sternum a tumour had been removed four months previously, which had been growing four months in a black mole. It was removed by Dr. Richards **of** Redruth, when about the size of **a** duck's egg, **and** was supposed **to have** been **a** cancer. When I saw the patient there was clearly a melanotic growth in **one** breast, and many disseminated tubercles of melanotic sarcoma scattered over the sternum, breasts, **abdomen,** and enlarged lymphatic axillary glands, for which no operation was justifiable.

In chapter xx., which deals with associated growths of different kinds, two cases, 8 and 9, will be found in which carcinoma of the breast and melanotic sarcoma were found in the same patient.

CHAPTER XII.

CARCINOMA OF THE BREAST
(AN EPITHELIAL INFILTRATING GROWTH).

CARCINOMA always originates in epithelial tissues, and is composed of epithelial elements.

These elements invariably *infiltrate* the tissue in which they are placed, and spread from their primary seat to neighbouring structures by a progressive infiltration, so that as the disease advances, skin and fat, muscles, bone, nerves, and vessels eventually become involved, although the structures named vary in their powers of resisting the infiltrating tendency.

In carcinoma of the breast, the gland structure is the one primarily involved ; and the epithelial elements of which it is composed partake of the anatomical characters of the normal epithelial lining of the ducts, or acini. The cells in the early stages of disease where they are in contact with the basement membrane of the gland are arranged with a certain kind of regularity; but at a later period they are most irregular, in the lumen of the duct, in the centre of the acinus, and in the connective tissue of the parts outside the basement membrane ; the vessels of the growth are distributed to the fibrous stroma of the tissue infiltrated, and not, as in connective tissue or sarcomatous tumours, between the cells.

The epithelial cell elements in their infiltrating

progress affect the invaded tissues in a particular way, and the cells in an early, if not in the earliest period of disease are found to be placed between the normal fibres, ducts, or gland elements of the invaded structure in what are **called** alveolar spaces. These **spaces** vary **in** shape, size, **and** in the **arrangements** of their epithelial contents, and **each alveolar** arrangement apparently depends, in the early period of disease, upon the anatomy of the tissue primarily infiltrated.

The alveolar spaces, whether large or small, freely communicate with one another in different places, and when a section of a tumour is made, and its surface scraped, the contents of the **spaces** readily escape **as** " milky" or " cream like" *cancer juice ;* this juice being composed of the epithelial cells, massed together, or floating in an albuminous fluid ; the cells composing the juice being large or **small,** with one, two, or more round or oval nuclei.

This alveolar arrangement does not, however, hold good for ever ; **it does** so only so long as any normal tissue remains uninvaded or unchanged. For it is to **be** recognised that in all cases of carcinoma, as the infiltration of the epithelial elements progresses, so the disappearance of the tissue infiltrated is to be traced, until at last the natural structure of the invaded parts, even of bone, becomes entirely replaced by the carcinomatous elements. Over and above this, there **is more or less** in all infiltrating tumours **a** proportion **of** granulation or indifferent cells, from which of **themselves** the nature of the tumour could **by no means** be distinguished.

Neoplasms, like **normal** tissues, are liable also to certain pathological processes, and tumours of the breast are obedient to this law ; thus they may inflame and suppurate, or ulcerate, and they may likewise undergo **chronic** degenerative changes. The life history of these epithelial or carcinomatous infiltrating tumours is not usually a long one, and every such tumour having arrived at maturity, begins to degenerate and to die, the process of death being in some cases slow, in others rapid. When slow, the process is by what is known as caseation; when rapid, by sloughing; this latter process, in some cases, being apparently helped by what much simulates an inflammation. At times a tumour may caseate in its centre and break down, so as to form a cavity containing a bloody or serous fluid simulating a cyst. In rarer cases the growth will undergo what is called the "colloid transformation," a form of degeneration peculiar to carcinoma, and not known to occur to the sarcomatous group. Of external tumours, those of the breast are mostly liable to this form of degeneration, whilst the more common examples of colloid transformation have been usually met with in the abdomen. In this change the growth becomes a gelatinous mass of a yellow or pinkish-red colour, and on section this material flows like transparent honey or mucus; it differs, however, from mucus in not being rendered opaque by acetic acid, and in containing sulphur; the alveoli of the changed growth enormously increase in size, and the epithelial cells of which it was composed swell into glistening globes, unaltered epithelial cells being often found in the centre of the mass.

In what way the epithelial cells of carcinoma that infiltrate the invaded tissue are formed and propagated may be open to dispute. In some cases it would appear that from either a known or unknown source of irritation, the natural elements of the epithelial tissue increase and multiply, and following the line of least resistance, dip into the tissues upon which they are placed, and so infiltrate them.

In others, where glandular structures are involved, as in the breast, the epithelial elements may first collect within the acinus, or tubule of the gland, and then make their way into the tissues outside, from rupture of the basement membrane of the acinus wall.

In a third class of cases, or perhaps in all to a degree, **the** disease spreads by what is called "*infection*," that is, the acquired power possessed by the developed morbid epithelial cells, when coming into contact with embryo undeveloped cells, of influencing their development, and causing them to take on the epithelial form. In the same way as in the cicatrisation of a granulating wound, the epithelial cells of the true skin at the margins of the sore influence the embryonic granulation tissue elements in contact with them to form new skin.

How far this "*infecting*" power **influences the** young connective tissue cells to assume the epithelial form in the growth of tumours, which originate in a glandular organ, and secondarily invade connective tissue structures, it is not possible to say; although it is probable that such an influence is an important one, since it is certain that epithelial growths increase rapidly **as** soon as their epithelial elements have escaped from the acini of the gland, in which they originally increased and multiplied, by what is known as ordinary generation, and become disseminated into the fibrous or connective tissue structure around the gland.

So long as a new formation in a glandular organ is confined within the limits of the organ; so long as it retains both in structure and function the type **of** the normal tissue, it constitutes an adenoma. But when the new growth "becomes destructive towards the surrounding tissues," when it proceeds to infect the neighbourhood, it becomes **cancerous.** The additional element in the case of **cancer** is clearly secondary to, and not co-ordinate **with, the** initial disturbance. Epithelial cells, and in fact the same kind of epithelial cells, are produced in two ways, but the two modes of formation do not belong to the same category; the one is a modification of healthy action, and the other is simply a mimicry of the original departure from the normal.

Carcinomatous tumours, moreover, at some time of their history cease to be local diseases; and it seems probable that this great change in the clinical character of the disease is marked by the pathological change, to which attention has just been drawn, namely, the escape of the epithelial elements through the basement membrane of the tube or acinus in which they primarily grew, and the secondary infection of the surrounding connective tissue by the epithelial elements.

"The cancerous element in the disorder," writes Creighton, "first shows itself when the infiltrated epithelium produces what we agree to call infection of the connective tissue cells with which they are in contact."

How cancer spreads.—When a cancer or carcinomatous growth ceases to be a local disease, it spreads in three marked ways: by "continuous or local infection," by "lymphatic infection," or by "secondary vascular infection"; and whilst in any given case one form of infection may be more marked than another, in others all forms may exist together. In the breast any one or all may co-exist.

By "*continuous local infection*" is meant the gradual involvement of surrounding structures, in the order of their arrangement, around the primary seat of the disease, by progressive infiltration, as well as by extension along the perivascular sheaths of the blood-vessels of the diseased part; this being a common feature of scirrhus.

By "*lymphatic infection*" is meant the infiltration of the lymphatic glands associated with the primary diseased centre, or its coverings, by the lymphatic ducts, the elements of the disease being carried either by the lymph to the glands, or the lymphatic ducts themselves becoming directly infiltrated.

By "*secondary or vascular infection*" is meant the propagation of the disease by other than the two methods already described, and probably by the blood currents; the evidence of this infection being afforded by the existence of one or more secondary growths, similar to the primary growth, in the viscera or other parts of the body, as, for example, in the pleura or liver in cases of scirrhus.

Local recurrence after removal of the primary growth in apparently the most complete manner is another sign of malignancy which is common to the carcinomata, as well as to some forms of sarcomata. It is a common feature of breast cancer.

With these general remarks on carcinoma, I will pass on to consider the disease as it affects the breast, and is seen at the bedside.

Signs and symptoms of carcinoma.— When a surgeon finds in the functionally inactive breast of a woman over forty years of age, or in one even ten years younger, a thickening or induration and apparent infiltration of one of its lobes, and with this local infiltration he fails to discover any other symptom or sign of a local inflammation, he is at once led to suspect the existence of a *carcinoma ;* and when in the course of a few weeks or months the infiltrated lobe has increased in size, or the infiltration has spread from the lobe of the gland in which it originated, either to other lobes, or to some of the tissues that cover or lie beneath the breast, the diagnosis of a scirrhous carcinoma, or a cancerous tumour of the breast is plainly confirmed ; the spreading of the local affection being indicated by either a "dimpling," or "puckering" (Plate III.), or

infiltration of the skin covering in the lobe, or by
diminished freedom of movement of the breast over
the pectoral muscle.

When, again, a pathologist makes a **section of a**
tumour involving the breast, and the section cuts
crisply and becomes concave, shows a glistening surface of a greyish-white colour, and dotted over with
yellow spots and streaks; when he finds the disease
to have originated in the gland structure, and to have
infiltrated its tissues as well as to be deficient in
everything like a capsule; when he scrapes the cut
surface, and the scraping yields a milky juice (cancer
juice), which under the microscope presents epithelial
cell elements, with large round or oval nuclei of great
variety ; and when, on a closer histological examination,
sections of the growth are made, and the pathologist
discovers that these epithelial elements have been
scraped from larger or smaller slit-like oval or rounded
alveolar spaces lying between the tissues of the affected
lobe or gland, and **that** these epithelial elements
otherwise possess no definite order of arrangement,
he at once pronounces the disease he is examining to
be carcinomatous or cancerous.

It must be understood, however, that there **are**
no such special cells as "cancer cells." Those
found in cancer are truly epithelial, but they
are enlarged and deformed, many of which possess
multiple nuclei, and are very prone to undergo fatty
degeneration. Their arrangement also differs widely
from the normal : they have no such regularity in
their arrangement upon their basement membrane as
seen in the adenomata, but their membrana propria is
destroyed, and the lymph spaces of the connective tissue
are infiltrated by the solid cell cylinders.

Should the pathologist, on further dissection, discover that the tissues outside the area of the gland
are involved in the disease; that no boundary line

PLATE III.

Fig. 1.

Fig. 2.

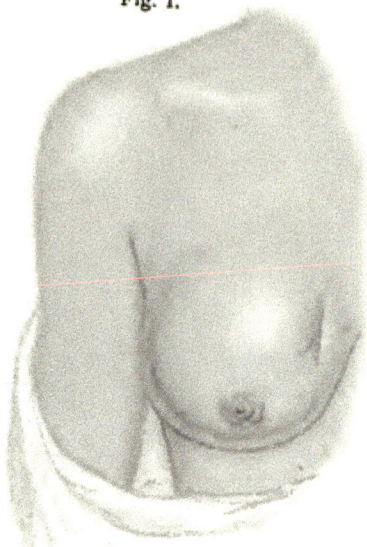 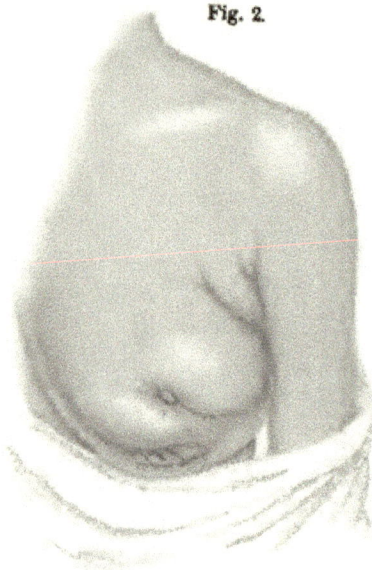

Fig. 3.

Fig. 4.

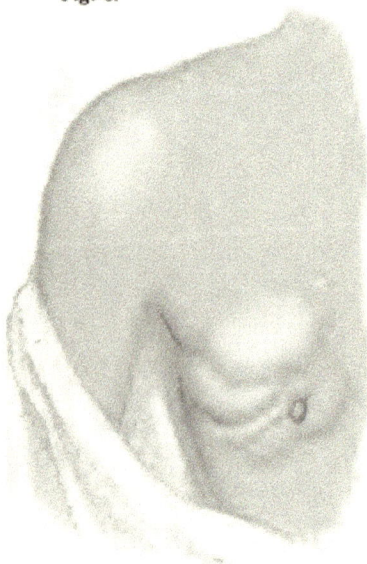 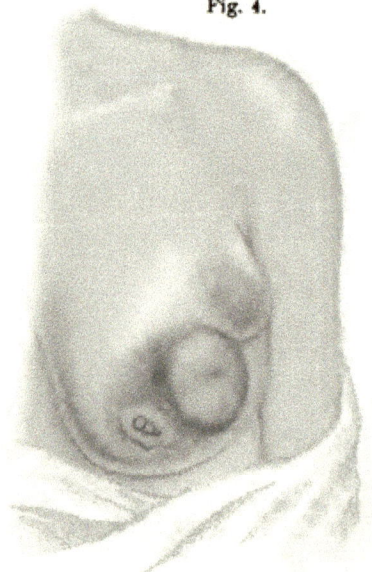

CARCINOMA OF THE BREAST.

1. Dimpling of Skin.   3. Atrophic Carcinoma.
2. Puckering of Skin.   4. Tuberous Carcinoma.

of any kind can be made out to exist between **the** healthy and diseased parts; that the capsule of the gland where it originally encased the healthy lobe has disappeared, and been replaced by a new cell growth ; and that this same form of neoplasm which originated in the centre of the diseased gland has radiated outwardly through the capsule of the gland into the fatty structure that surrounded it, and probably through this fatty structure either to the skin or the muscles beneath, so as in the former case to bring about, first, a drawing downwards of this structure (dimpling, Plate III. Fig. 1), secondly, early infiltration (puckering, Plate III. Fig. 2), and thirdly, marked infiltration (induration), or in the latter case fixedness of the growth and its immobility from the pectoral muscle : when these facts **are** discovered the diagnosis of cancer is absolute. Indeed, both **the** clinical as well as pathological facts demonstrate that **the** disease, which began by infiltrating a lobe of the **gland,** has spread rapidly by the same infiltrating process ; that it has passed from the lobe primarily involved through its capsule, and from this through adipose and connective tissue, either to the skin itself or pectoral muscle beneath, so that at last the skin had become infiltrated, and so changed. For carcinomatous infiltration of any tissue means eventually epithelial substitution ; and tissues that are at first invaded or infiltrated, in the course of time entirely disappear, their elementary tissues steadily giving way under the infective influence of epithelial infiltration ; so that at length the natural structure of the mammary gland becomes superseded by the epithelial elements.

From these considerations it may be asserted that four main pathological points so far stand prominently forward in carcinoma of the breast. The *first* being that **the disease is** an infiltrating one ; the *second,* that the infiltrating elements are epithelial, and nothing else ; the *third*, that whilst the disease may have originated in one structure, say one lobe of the breast, it will eventually locally infect neighbouring structures by a progressive infiltration (extension by local infection) ; and *fourthly*, that the normal structures invaded or infiltrated will eventually be destroyed and superseded by the epithelial infiltrating material.

Cancerous tumours, however, whilst they locally spread by *local infection*, do so in other ways, and of these, the method by *lymphatic infection* is the most frequent, the epithelial elements being either carried directly by the lymph stream from the lymphatic spaces of the infiltrated tissue, through the lymphatic ducts, to the neighbouring axillary, clavicular, or substernal glands, or the walls of the lymphatic ducts themselves become absolutely infiltrated with the cancerous elements, and thus the local disease is conveyed by direct extension to the lymphatic glands themselves. It is probable that both these two methods have their influence, and that the first is the more common.

Again, carcinomatous disease spreads by "vascular infection," that is, by the blood-vessels, and so becomes disseminated broadly into the viscera and other distant parts of the body ; local infection, lymphatic infection, and vascular infection being the chief methods by which cancerous disease spreads.

The varieties and macroscopical features of carcinomatous growths. — There are clinically six classes of cases of carcinoma of the breast, although "the varieties of carcinoma are determined by the relative proportion of the stroma and cells, by certain degenerations and transformations, and by the accidental formation of cysts " (Gross).

1. The very slowly growing, and subsequently

atrophying, scirrhous carcinoma (Plate **III.** Fig. **3 ;**
Plate **V.** Fig. 3 ; Plate VI. Fig. 1).

2. The more common variety, or hard, fibrous, infiltrating form of carcinoma (Plate V. Fig. 2).

3. The soft, more cellular, and encephaloid variety (Plate **V.** Fig. 4).

4. The acute brawny cancer (Plate IV. Fig. 3).

5. The colloid cancer (Plate VI. Figs. 3 and 4).

6. **The** cystic carcinoma (Plate VII. Fig. 4).

The *first* variety (Plate III. Fig. 3), or the very slowly growing, and subsequently atrophying, form of scirrhus, is the hardest kind met with, and is often described **as** stone cancer ; it rarely attains any large dimensions, and is characterised by its peculiar power of contracting all the tissues involved in its infiltrating influence into little more than a puckered scar, with a central hard stony **nucleus.** When the centre of **the** breast is its starting point, the breast gland, with **its** nipple and skin covering, may **all** be drawn together into a cicatrix-like fissure. And **when** the primary growth occupies the periphery of the breast gland, the **same** contractive powers will be visible, although without any retraction of the nipple. In Plate VI. Figs. 1 and **2** this condition is well illustrated.

The disease may progress so slowly as **to** last twelve, **or** even twenty years, and at the end appear **only as a** local disease ; whilst in other cases a tumour, **that for** years had grown but slowly, may suddenly **take on** active growth, and develop into an acute or subacute form of **cancer.**

During the **progress of this** withering scirrhus, tubercles may appear **in the skin** of the primary growth and in its neighbourhood, which may appear and **even** disappear, the disease progressing and receding **at the** same time. Should the growth be irritated or **interfered with, it** may become active, and what has been a local disease will become a general one by lymphatic and vascular infection. Patients the subject of this form of cancer, as a rule, die from visceral metastatic growths rather than from the local disease. A section of a tumour, such as the slowest growing of this series, will cut crisply, present a concave surface of grey appearance with the remains of old gland ducts. There will exude from its surface but little juice, and this juice will contain but few epithelial cells ; and what are seen will probably be undergoing fatty or granular degenerative changes. The fibrous stroma will be abundant, and the alveoli which contain the epithelial elements will probably only show as slits or fusiform clefts. This form of carcinoma may clinically for a long period be regarded as a local disease, although at a later period it spreads by lymphatic infection.

In the following brief notes of cases, many of these points will be well illustrated.

1. *Carcinoma fibrosum of right breast ; disappearance of local disease by natural processes ; death of patient of chest disease seventeen years after first appearance of disease.*—H. B., a healthy looking, childless, married woman, aged 53, came under my care in January, 1857, with an ulcerating carcinomatous tumour of her right breast. The disease had existed for six years, and ulceration had been present for four. The cancerous surface was about the size of the palm of the hand. It was of stony hardness and firmly fixed to the muscles. Its edges were nodular and crumbly ; the axillary glands were also enlarged. Her general health was good, and as the tumour caused little pain, operative interference was rejected. In March, 1858, fifteen months later, the tumour had become much smaller, and several pieces the size of nuts had fallen off, having apparently been destroyed by the contraction of the fibrous

elements of the neoplasm. The growth as a whole
was much harder. General health still good. October
20, 1858, six months later, much of the original
tumour had crumbled away, and the tumour was
much smaller. Some tubercles had, however, **ap-**
peared **in the skin** over the sternum.

April 21, 1859. Tumour continues **to** contract,
and to throw off pieces. The tubercles in the skin
are likewise contracting and becoming paler.

November, 1859. Axillary glands becoming
smaller and more indurated.

July 23, 1861. The breast has nearly cicatrised,
a mere linear puckered scar remaining, in which are
one or two small hard white tubercles. All the
secondary tubercles of the integument have disap-
peared.

January 3, 1862. Nearly cured.

June 1, 1862. Only one small tubercle the size of
half a nut remains in the cicatrix. No fresh tubercles have appeared. The glands in the axilla can
hardly be felt.

March **31, 1863.** Breast shows merely a cicatrix,
in which there **is only** one small nodule the size of a
pea. The woman in all other respects is well.

May 30, 1864. The breast is still in the same
condition ; one or two tubercles have appeared in the
integument near the cicatrix, and have again disappeared. Two tubercles are, however, still present.

June 30, 1864. The patient considers herself to
be **well.** **Her** skin tubercles cause no pain or incon-
venience. **This is the** last note of her case I have,
but I learnt later that this patient died in 1870 of
some chest trouble. **The** local disease had **not**
increased.

2. *Atrophic cancer, twelve years' growth ; patient
survived fifteen years.*—Miss B., aged 50, who had had
atrophic cancer of her left breast for twelve years,

with enlarged axillary glands, came under my care on September 13, 1868, when the nipple and skin around, and breast gland were all involved in a hard, puckered, cancerous tumour. This patient lived for three years, and died from some acute chest trouble.

3. *Atrophic cancer of twenty years' standing.*—Jane H., aged 72, the mother of two children, and a widow of fifteen years, came under my care in 1869 with a puckered carcinomatous infiltration of her left breast of twenty years' existence. The whole gland, and the skin over it, and the axillary glands were involved, although the nipple was natural. Secondary tubercles were very general in the skin. This patient died one year later of asthenia.

4. *Atrophic cancer of breast, nine years.*—Bridget M., aged 56, the mother of three children, came to me on August 18, 1868, with an atrophic cancer of her left breast, which had existed for nine years, with skin tubercles for the last two years over the breast and surrounding skin. She was steadily losing strength, and died within a year.

5. *Atrophic cancer of breast, eight years.*—Ann H., aged 69, the mother of one child twenty-eight years old, came to me on May 31, 1866, with infiltrating carcinoma of the left breast of seven years' growth. The nipple was retracted and the axillary glands were enlarged. The skin over the breast was only puckered, not infiltrated. This patient, I believe, lived three or four years after my seeing her. The cause of her death is uncertain.

Second variety (Plate III. Figs. 1 and 2; Plate V. Fig. 2).—The second or more common variety of the hard, fibrous, infiltrating form of cancer differs only from the last, or atrophying kind, in that it is more rapid in its progress, and does not tend to wither. In it the bundles of fibrous tissue are abundant, and the cell elements comparatively limited;

at the same time they are far more abundant than
they are in the atrophying kind. The alveolar spaces
which contain the irregularly heaped-up cells are
ovoid or round, and the disease runs its course far
more rapidly; three years being about the average
duration of life of a person who is the subject of it.
This **variety of** cancer is the **common form, as met
with in practice.**

The **tumour rarely** attains a size larger than that
of an egg, but it **never** retains the smoothness and
regularity of its shape. It has always an irregular
bossed outline, and may affect the skin in ways to be
described (chapter xiii.). A description of a section of
this tumour has been given (page 138), the tumour,
taken as a type of carcinoma of the breast, having been
drawn from this group, **since** it is **the** most common.

Third variety (Plate **III.** Fig. **4**; Plate **V.**
Fig. 4).—The soft tuberous, **or** encephaloid variety
of carcinoma is likewise an infiltrating one, but it
differs from **the two** preceding varieties in being
larger in size, more rapid in its growth, and softer in
its consistence. It may readily be broken down by
the finger. On section it has a homogeneous, brain-
like, **white or mottled** pink **or** red surface, with probably some more or less extensive local extravasation
of blood within its structure. **In** parts, the tumour
may be breaking down from degenerative processes.
The cut surfaces are less concave than are the **scirrhous forms, and** much juice exudes **from them. The
juice contains** abundant cell elements ; **a fine section**
of the tumour **shows little** fibre tissue, but **large,
round, alveolar spaces filled with** round large cells.

4. *Fourth* variety (Plate **IV.** Fig. 3).—The acute
brawny cancer, **although** it simulates the two last
varieties, is a variety by itself, in that it is an acute
disease, and appears as a rapid infiltration of a breast
with the skin over it. The nipple, **as** a rule, is

K—25

retracted and much depressed; the soft parts around
it are raised, and the nipple depression is thus made
apparently greater. Occasionally, as in Plate IV
Fig. 3, the nipple is infiltrated and raised; the skin
over the breast is brawny with infiltration, and at times
it may be œdematous, so as to pit on pressure; it may
likewise be injected as if inflamed, and hot to the hand.

In other cases the skin over the infiltrated breast
gland will be indurated and tuberculated, the skin
and the tumour having become by local infection one
infiltrated carcinomatous mass.

A section of such a tumour will be firm, but not
so hard as in class 2, or so soft as in class 3; it will
present on section a white glistening appearance, with
abundance of cancer juice containing cell elements
(Plate V. Fig. 4). The fibre tissue will not be very
abundant.

To the pathologist's eye the disease will probably
be able to be traced from the gland or lobe in which
it originated into the neighbouring structures, and
even into the skin, the section of which will be thick
and white, from the infiltration of the epithelial cell
elements.

This form of cancer of the breast is the most acute
met with, and the most rapidly fatal; it may run its
course in a few months. It is a type of an acute
infiltrating neoplasm.

5. *Fifth* variety.—**Colloid carcinoma** (Plate
VI. Figs. 3 and 4).—This is a degeneration of a carcinomatous growth to which attention has been drawn
(page 134, and also at page 199).

6. *Sixth* variety (Plate VII. Fig. 4).—The cystic
carcinoma will be considered under the heading of
cystic disease of the breast. The existence of cysts
in the carcinomata clinically separating cases which
contain them from the more solid varieties of the
disease, in the same way as the presence of cysts

renders a different consideration necessary of the cystic and solid forms of adeno-fibromata and sarcomata.

Of the four first classes just described, there are endless intermediate varieties ; and there is a chain of connection which links at one end, class 1, as represented by a small tumour of stony hardness, which has slowly drawn to itself, by local infection and infiltration, breast tissue, with its coverings, so as to appear as an indurated, puckered, and painless nodule, with classes 3 and 4 at the other end, which exists as a soft, rapidly growing, succulent tumour, the size of a fist or cocoanut, having a rounded nodular outline and elastic feel, with full veins coursing over the skin which covers it, or with the skin mottled, infiltrated, and consequently adherent to the parts beneath.

In the former case the growth will have been of slow growth, even of many, say twenty, years ; in the latter case it will have been of rapid growth, probably of weeks.

Neither of these extreme types are, however, of frequent occurrence, as one represents the withering or atrophying form of scirrhus, and the other the exceptional encephaloid or tuberous kind.

Influence of sex.—Carcinoma of the breast is met with in both sexes. Mr. W. R. Williams, in his work on "The Influence of Sex in Disease" (1885), tells us that from an analysis of all cases admitted into the Middlesex, St. Bartholomew's, St. Thomas's, and University College Hospitals, during ten, twelve, and seventeen years, ending 1883, "out of 11,100 cases of tumour, 53 per cent. were cancerous ; out of 5,978 cases of cancer, 24 per cent. were of the breast, and out of 1,433 cases of cancer of the breast consecutively observed in the above hospitals, 14 were in males.

Two out of every five cases of cancer in women are of the breast ; one out of every three are of the uterus.

There is but one case of cancer of the male breast to 101 of the female breast.

Out of one hundred consecutive cases of carcinoma of the breast which I have observed, six have been of the atrophying form (class 1), ninety of the ordinary scirrhous variety (class 2), and four of the brawny kind (class 4).

What has been usually described as medullary or encephaloid cancer I now believe to have been probably of a sarcomatous nature; although some of the cases may have been of the tuberous variety, class 3. The fact of a tumour being encephaloid and limited by a membrane clearly excludes it from the carcinomatous series, carcinoma under all circumstances being an infiltrating *neoplasm.* Paget believes that there are about five cases of encephaloid carcinoma tc ninety-five of scirrhus; whilst Lebert states that in France one case of encephaloid is met with in every five cases of carcinoma.

Influence of age.—Carcinoma is not a disease of young life, although Mr. Lyford has recorded* a case in which it was found in the breast of a girl, aged 8; Mr. B. B. Cooper has reported a second † in a girl, aged 16, and in the Museum of St. Bartholomew's Hospital there is a specimen which was taken from a girl, aged 16.‡ Dr. Henry, of Breslau, recorded a fourth case in a woman, aged 21; and Mr. Tilley a fifth in a woman, aged 26. I have seen one example in a woman, aged 25, and two examples in women, aged 28, one of which was of the scirrhous variety, and the second of the acute brawny kind. Within the last few weeks I have seen a case in a young married woman, aged 26, in whom the disease appeared during pregnancy. It had all the features of the brawny variety. Nothing could be done for it.

* *Lancet,* vol. xii. p. 332.   † Lectures on Surgery. 1851.
‡ Prep. Series xxxiv. 4.

An analysis of 600 consecutive cases of my own tells me that

Years of age.

In 25 cases, or 4 p. c., the disease was discovered before 30

163	,,	27	,,	,,	,,	,,	between 31 and 40
216	,,	36	,,	,,	,,	,,	,, 41 ,, 50
150	,,	25	,,	,,	,,	,,	,, 51 ,, 60
44	,,	7	,,	,,	,,	,,	,, 61 ,, 70

2 cases the patients were over 70 years of age. .

The statistics of other authors are very similar to these. Thus Sir J. Paget found that of 400 cases, 98 appeared before the age of 40, or 24·5 per cent., and 302, or 75·5 per cent., after that age.

From these facts it is clear that only three out of every ten cases of carcinoma attack the breasts of women during the period of its functional activity, that is, before the age of 40 ; seven cases out of every ten occur in the breast gland during the years of its functional decline, that is, after the age of 40; the decennial period between the ages of 41 and 50 being the one in which the disease is the most prone to appear. The conclusion is clear that in exceptional cases the disease may appear early or late. Independently of the cases analysed I have seen one which appeared at the age of 96, in the breast of a lady who had had many children, and died from old age ; and a second which was discovered at the age of 86. Mr. Ashhurst saw one at the age of 89.

Social condition and fecundity of patients. —Of the 600 women who had carcinoma, 485, or 80 per cent. of the whole number, had been or were married ; 115, or 20 per cent., were single. Baker's statistics are about the same : 77 per cent. were or had been married, and 23 per cent. single. Gross, on the other hand, out of 688 cases, found the married bore the proportion to the single of 88 to about 12 per cent., the proportion of single being less than my own tables show.

Of the 485 women in my own table who were or had been married, 360 were prolific, or 74 per cent., and 125 sterile, or 26 per cent. A large proportion of the prolific women were so to an extreme degree, ten and more children to one mother being a common note to find recorded. Winiwarter has noticed this same point, six children in his cases being frequently recorded to one mother. The breasts of married women and of those in whom the gland has been the most active are apparently more liable to cancer when the period of gland activity has passed than are the breasts of single women. A gland that has been functionally active being more prone in its period of obsolescence to become the seat of carcinoma than another which has never been called into activity.

The above facts show how erroneous is the common assertion that the unmarried women are more liable to carcinoma of the breast than the married.

Breast involved.—In 300, or half of the 600 cases, the right gland was the seat of disease ; in 272 the left ; and in 28 both glands were involved ; double cancer being apparently present in 5 per cent. of all cases, or in 1 in every 20.

Influence of menstruation.—My own facts are not clear enough upon this point, but Birkett, who is known to be a very careful observer, states, " that in a large majority of the women who have cancer of the breast the function was persistent at the moment of the development of the disease." That is, in 70 per cent. of his cases the catamenia were persistent, and in 30 per cent. it had ceased. Gross states that in his cases 61 per cent. of the women were menstruating, and in only 6·4 per cent. of them was there irregularity in the performance of that function. In 39 per cent. the catamenia had ceased.

Winiwarter does not believe that menstruation has

any ætiological connection with carcinoma, and in this
I quite agree.

Duration of disease when first seen.—This
point was noted in 504 of the 600 cases.

In 221 the disease had existed under a year.
  ,, 165 it had been discovered between 1 and 2 **years.**
  ,, 39 it had existed between 2 and 3 years.
  ,, 25 ,, ,, ,,   3 ,, 4 ,,
  ,, 22 ,, ,, ,,   4 ,, 5 ,, } 15 per cent. over
  ,, 15 ,, ,, ,,   6 ,, 7 ,,   3 years.
  ,, 10 ,, ,, ,,   8 ,, 9 ,,
  ,, 7 it had existed from 10 to 20 **years.**

From these facts it is clear that in three out of four
cases of carcinoma of the breast the disease when first
seen by the surgeon has usually existed for less than
two years. In the fourth case it has existed from two
up to twenty years. In the majority of these latter
cases, and in all those that had given a history of
more than four years' duration, the disease was probably of the atrophic kind. I should not, however,
have believed, if I had not made the above analysis,
that so large a proportion of the cases of carcinoma of
the breast of which I had taken notes had been of so
chronic a nature as these facts indicate, for it is to be
remembered that these figures give the duration of the
disease before treatment, and if the benefit of treatment is added, a considerable extension of time would
probably have to be given.

If we accept Paget's conclusion, that the average
duration of life in carcinoma of the breast is about
four years, and my own cases confirm this view when
thus divided, we must accept as equally true a more
important and hopeful conclusion, that a large number
of patients, approaching a third of the whole number,
survive this period for months or years. This conclusion will be very palpable on looking at the following

table, composed of an analysis of 72 operation cases, the results of which I have been able to trace.

Table showing the duration of life of 72 cases after the carcinomatous tumour had been discovered, all of which had been operated upon with success.

In 8 or 11 p. c. the disease ran its course within 12 months.

„ 16 „ 22 p.c. patient died between 12 and 24 months

„ 16 „ 22 „   „   „   „   24 „ 36   „

„ 10 „ 14 „   „   „   „   3 „ 4 years.

„ 2 „ 2·7 „   „   „ in the fifth year.

„ 8 „ 11 „   „   ,,   between 5½ and 6 years.

„ 2 „ 2·7 „   ,,   lived eight years.

44 per cent. 55 p.c. died within over three years. three years.

2 lived nine and 6 ten years, or 14 per cent. over six years.

55 per cent. of the whole number sank within three years, and 44 per cent. over three years; these proportions being precisely those brought out by Sir J. Paget in his analysis of cases which had not been interfered with. But my table tells us likewise that 18 out of the 72 cases, or at least one-fourth of all cases operated upon, lived from 5 to 10 years.

It is well, therefore, to have this fact always before us, and that whilst we assert in all truth that one half of our patients will die within the three years, we can with equal justice lead any individual patient to hope that her chances are equally good to live from 5 to 10 years. Indeed, we may with a clear conscience admit, that whilst one-third of all cases die within two years, two-thirds will live from 3 to 10 years; half of these surviving from 5 to 10 years.

The surgeon, with these facts before him, is therefore justified in placing before any individual patient the same hopeful aspect of her case, and thus giving encouragement where so much is needed.

I quote Paget's table of 61 cases, the records of which were complete, since it deals with cases which

have not been interfered with, whereas my **table** includes only cases after operation.

```
 7 died between  6 and 12 months. ⎫
 7    ,,      ,,  12  ,, 18    ,,   ⎪  34 cases under 3 years,
 8    ,,      ,,  18  ,, 24    ,,   ⎬      or 55 per cent.
10    ,,      ,,   2  ,, 2½ years.  ⎪
 2    ,,      ,,  2½  ,,  3    ,,   ⎭
12    ,,      ,,   3  ,,  4    ,,   ⎫
 6    ,,      ,,   4  ,,  6    ,,   ⎪  27 cases over 3 years,
 3    ,,      ,,   6  ,,  8    ,,   ⎬     or 46·2 per cent.
 1    ,,      ,,   8  ,, 10    ,,   ⎪
 5    ,,      ,,  10  ,, 20    ,,   ⎭
```

Baker gives the average duration of life 43 months; Sibley makes it 32 months; Gross only 27 months without operation, and 39 months with.

Carcinoma of the breast complicated with pregnancy.—Should the breast of a pregnant woman be the seat of cancer, the disease as a rule will progress rapidly, and should the stage of suckling be reached, its increase will be still more rapid. These cases are happily rare. I record a few examples.

Mr. Annandale **reports** that **he** once operated upon **a mother** and daughter for scirrhus of the breast, both within two weeks; the daughter was nursing **when** the disease first showed itself, and in her case its **progress** was much more rapid than in the case of her mother; and further, the disease returned in the daughter's breast much earlier than in the mother's.

1. *Extensive cancer of breasts and integument of the chest and abdomen, associated with pregnancy following suppuration of breast; death in eighth month of pregnancy.*—Mrs. M., aged 35, the mother of nine children, and the first eight of which she suckled, came to me in 1868, when she was four months pregnant, with the *right* breast and skin over it, **as** well as the integument of the chest, side, and **abdomen**, generally infiltrated with carcinoma. In

fact, she was skin-bound. The breast with the skin
in several parts was ulcerated. The right axillary
glands were enlarged, and the right arm œdematous
from venous obstruction. The disease in the *right*
breast had appeared as a lump in the gland, which
was left after an abscess that formed when suckling
her eighth child two years previously. The left breast
was likewise infiltrated with carcinoma, and this had
followed an abscess after the ninth child, ten months
previously.

This patient died from asthenia about the eighth
month of her pregnancy.

2. *Cancer of breast, followed by pregnancy, and miscarriage; after which rapid growth of disease; death.*
—Mary L., aged 44, the mother of four children, all of
whom she suckled, came under my care on January
19th, 1865, with carcinoma of her left breast. The
disease had been coming *for two years*, and had increased slowly; seven months ago she became pregnant, since then the disease has progressed more
rapidly; she miscarried at the fifth month, that is,
seven weeks ago.

At present the mammary gland as a whole is
infiltrated, and fixed to the pectoral muscle; the skin
over the gland is fixed, puckered, infiltrated, and
ulcerated around the gland; in the skin are many small
tubercles of cancer; the axillary glands are slightly
enlarged. Later on the ulceration spread fast, and
the tubercles multiplied over the skin and on the right
breast. In September the patient died with chest
complication. Her *paternal* grandmother had died,
aged 87, with chronic cancer of the breast; her *paternal*
aunt also, with the same disease, aged 65.

3. *Cancer of breast coming on during pregnancy;
acute progress involving both glands; early death.*—
Eliza S., aged 49, the mother of three children, whom
she did not suckle, the youngest being *four months*

old, came to me on March 18th, 1869, with a lump in her right breast, which she had discovered three months before her last confinement, and which had rapidly increased. When **seen seven** months after its appearance, the whole gland was infiltrated with disease and the skin **over it was** brawny. The tumour was fixed to **the chest;** the axillary glands enlarged. There **was** pain **down** the arm, which was œdematous.

Four months later the disease had affected the opposite **breast,** and the woman soon sank. The patient's mother had died from tumour, and her sister from cancer of the breast.

4. *Cancer of breast with retracted nipple, which prevented suckling, following labour.*—Mary B., aged 38, the mother of five children, the youngest being five months. She **had** suckled all with her right breast, **but not** with the *left,* as the nipple was retracted.

November 4th, 1869. The left breast was infiltrated with cancer, associated with enlarged axillary glands. It had commenced after her last confinement, five months previously. Six months later this woman died.

5. *Acute brawny cancer of breast, appearing during lactation.*—Eliza W., aged 35, the mother of four children, the youngest being fifteen months, all of whom she had suckled, came under my care on January, 1871, with infiltration of the *left* breast and a brawny condition of skin over it. The disease **began** three months before, during suckling, as a lump which rapidly **grew.** The skin over the right breast was mottled **and was** already becoming infiltrated in lines, probably **lymphatic.** On March 2nd the woman was sinking. **Her** mother's cousin had cancer, and two cousins, **one paternal,** and one maternal, had tumours.

General condition of patient the subject of carcinoma.—An old prejudice still exists in favour of the presence of a cachexia in cases of cancer ; although a very little clinical experience is wanted to learn that this prejudice is wrong, and that there are few patients admitted into metropolitan hospitals with a more general healthy aspect than those who have cancer of the breast. At least half the cases who have this disease present the appearance of perfect health. Two-thirds of the remainder look as healthy as the bulk of those who are doing the work of life well. Whilst in the sixth, not included in either of these classes, some general evidence of illness may be made out ; the general evidence depending more upon mental anxiety and apprehension than any real influence of the disease upon the functions of the body. To see what is known as cachexia, the surgeon must go into the cancer wards of a hospital, in which the sufferers from cancer are retained until the end, that is, until life is slowly sapped by hæmorrhage, discharges, pain, and the interference with some one or more of the important functions of the viscera from metastatic or secondary growths. These patients look ill and cachectic, it is true, but probably not more so than others who are suffering in an allied way, though not from cancerous affections. The cachexia of cancer, in fact, differs in no way from that of any other exhausting disease. When it exists it indicates the presence of some affection which is undermining the patient's strength, which may be cancer, but it may be any other form of disease.

The observations of other writers support these conclusions, for Gross states that fifty-one per cent. of his cases were in good health, nineteen per cent. were pale and thin, and twelve per cent. were decidedly broken down from the effects of the disease. He adds also that the nutrition of scarcely one in twenty suffers previous to sixteen months after the detection of the growth. Paget found that sixty-six out of

ninety-one cases presented the characters of good
health ; nine were in but moderate health, and only
sixteen were sickly.

Heredity in cancer.—Out of my 600 cases **of**
cancer, although carefully inquired into, I found it to
exist **in** only seventy-three cases, or 12 per cent. of the
whole **num**ber.   In fifty-four of these, cancer was
reported to have occurred in one member ; **in sixteen
in two members, and in** three in three members.   **In
fourteen of** the seventy-three cases the relation was
on the father's side ; in thirty-five on the mother's
side, and **in** eleven it was found among the brothers
and sisters. In seven of the thirteen remaining cases it
had attacked the aunts, in five the cousins, and in one
the grandmother, my notes not stating whether
maternal **or** paternal relations.   These facts **are
enough** to indicate that an hereditary tendency **to**
cancer of the breast is by no means the rule (only
12 per cent.), and that where it exists it seems to be
more powerful on the mother's than the father's
side.

Gross found a history **of** heredity in but 9·7 per
cent. ; **Paget in 33 per cent. ;** Nunn in 29 per cent.
Birkett is **unable to** support **the** argument of heredity.   Nunn has, however, pointed out **a** fact of
interest, and that is the longevity of the families of
patients who have cancer of the breast.   He found in
169 cases, that the average age of the fathers was
sixty-two, **and** the average age of the mothers was
sixty-one.   **That** 106 of the patients had had grand-
parents who **had** lived over seventy years of age ;
sixty-two had parents who had passed the great age
of eighty, and fifteen who had lived over ninety years
of age ; these facts neither suggest that cancerous
patients come from a feeble stock, nor that they in-
herit the affection, but rather that the disease is an
acquired and a personal one.

I may quote here a rare case of carcinoma which existed in mother and daughter at the same time.

Carcinoma of the breast in mother and daughter; of slow growth in the mother, acute in the daughter; in the latter it had followed a blow.—In 1878 I saw Mrs. S., who was then 68, with carcinoma of the cicatrix of a wound which had been made thirty years previously for the removal of a cancerous breast of five years' growth. She had remained well after the operation for twenty-five years, when a return took place. When I saw her there was a tubercle the size of a walnut in the scar, with many other tubercles scattered in the skin over the sternum and opposite breast. Nothing could be done for her. She, however, brought her daughter to me, who was 50 years of age, with a decided carcinomatous tuber the size of a small orange in her breast, which had followed within a few weeks of a blow. The breast was wholly infiltrated, with the skin over it, as were also the lymphatic glands of the axilla and above the clavicle. Nothing could be done for this trouble, and the patient died within six months from probably internal thoracic cancer.

" When," says Darwin, "amongst individuals, apparently exposed to the same conditions, any very rare deviation, due to some extraordinary combination of circumstances, appears in the parent (say once amongst several million individuals), and it re-appears in the child, the mere doctrine of chances almost compels us to attribute its re-appearance to inheritance. Every one must have heard of cases of albinism, prickly skin, hairy bodies, etc., appearing in several members of the same family. If strange and rare deviations of structure are really inherited, less strange and common deviations may be freely admitted to be inheritable. Perhaps the correct way of viewing the whole subject would be to look at the inheritance of any character whatever as the rule, and non-inheritance as the anomaly."

If Paget's statement be true, as I take it to be, "that a mark once made on a particle of blood or tissue, is not for years effaced from its successors," why should we wonder at inherited tendencies in form and feature, in physiological function, and in pathological processes? why, indeed, should we not rather expect these inherited proclivities of tissues to have a greater effect, and look for more marked evidence of heredity in disease than we find to be the rule in practice.

The inheritance of a predisposition to disease is a very different thing to the inheritance of a disease itself. The former need never be made manifest; indeed, it is probable that, if it exists, it can be made to perish by the careful exclusion of all such influences as may bring it to light.

Causes of carcinoma of the breast.—All evidence tends to show that carcinoma of the breast, as of other parts, is in its origin more a local than a constitutional disease, and that it becomes a general one in a secondary way by what has been described as "local infection," "lymphatic infection," and "vascular or secondary infection."

To start in the breast, as elsewhere, some local source of irritation is required, and that local source may come from without in the form of an *injury;* or from within, either as some spurious functional activity, or degenerating involution change, or as some antecedent inflammatory change, which as an acute affection may have damaged, or as a chronic one may have irritated tissue.

It is now a recognised fact that persistent irritation of the skin may give rise to an epithelioma, and daily experience supports this view, particularly in disease of the lip, tongue, and penis. The growth

of an epithelioma in the bladder of a man, who for
years relieved himself by the frequent passage of a
metallic catheter through a perinæal fistula, at a spot
against which the catheter impinged, is an interesting
illustration of this view. The irritation of soot as
a cause of sweep's cancer, and the irritation of
common warts, are more familiar examples. The
fact also that cancer primarily is most prone to attack
the mucous canals at their narrowest parts, where
friction is most felt, supports this view.

Persistent local irritation of an epithelial structure
is consequently likely to produce a local cancer, and
when this local irritation takes place in a gland, such
as the breast (which is undergoing involution changes
and becoming obsolete) the tendency to produce a
carcinomatous tumour is increased ; when, in addition
to these local causes, there is added the predisposing
influence of heredity, by which the members of a
family tend to resemble not only their immediate
parents, but to comprise in themselves the foibles
of several generations, the pernicious influence of a
local irritation is likely to become rampant.

That new tissues, such as cicatrices, are prone to
become the seat of epithelioma, is well recognised ;
there is, indeed, no denying the fact that the cicatrices of burns, of old ulcers and stumps, are con-
stantly the seats of cancerous disease, and more
particularly when irritated ; we need not, therefore, be
surprised that epithelial gland structures, such as the
breast, should occasionally become the seat of epithelial growths, after having been injured by some
chronic or even suppurative inflammation ; the carcinomatous disease not following at once the local
affection which first injured the breast tissue, but
attacking it, as it might a scar, at a more advanced
period of life, the reparative material in the gland
and in the scar being both new, and in a measure

cicatricial. I have seen a cancerous tumour originate in the scar of an antecedent breast abscess, and many carcinomatous tumours have followed a previous mastitis.

Gross states that carcinomatous disease started from lumps, or chronic indurations left by puerperal mastitis in 8·21 per cent. of the 365 women who had borne children ; the lumps left by the mastitis having remained quiescent for a period which varied from four to twenty-eight years. My own statistics tell me that out of my 360 cases of women who had borne children, mastitis had occurred at some antecedent period in eighty, but it is almost impossible to make out that the cancerous tumour originated at the seat of scar or induration. The tumour in the majority of instances appeared many years after the mastitis ; in one case thirty years.

In the following case carcinoma clearly originated in the scar of an operation made twelve years previously for a cystic non-cancerous tumour :

Cystic degeneration of the breast, with enlargement of one of the cysts, and the development of an intra-cystic adeno-fibromatous growth ; removal of the growth ; twelve years later carcinoma of the cicatrix.— A healthy widow, aged 30, with one child ten years of age, consulted Mr. Birkett in October, 1859, for a small painless swelling, situated in the clavicular lobe of her left breast, which had been growing for seven months. The tumour was in a measure pear-shaped, with its narrow end near the nipple, nodular, and elastic. Pressure upon it caused a few drops of a clear, yellow, viscid, albuminous fluid, to exude from the nipple, which was natural. The axillary glands were not enlarged. In December, 1859, the tumour was excised with some of the breast gland, and the patient recovered. On section the tumour was found to be formed of a cyst, which contained

L—25

serum, and communicated with a duct near the
nipple.   Throughout the tissue of the lobe removed,
and of the gland structure exposed by the incision, were
small cysts, containing intracystic growths.   There
was no distinct, well-marked, limited envelope around
the tumour, but the cyst and all the new growths
were so closely identified with the ordinary tissue of
the breast, that it was absolutely necessary to cut its
structure in order to remove effectually the whole of
the tumour."   The intracystic growths were composed
of the cæcal terminations of gland tissue and fibre
elements, the former loaded with epithelium.

The patient remained well for eleven and a half
years, when a carcinomatous tubercle appeared in the
scar, which enlarged and ulcerated.   In June, 1872,
it formed an excavated ulcer, with thick everted
edges one inch and a half in diameter; the axillary
glands were enlarged.   October 15, 1872: the breast
and growth were removed, the patient recovering.

Direct injury is, without doubt, a common cause
of tumour formation, although not particularly of
carcinoma.   To say in what proportion such cases
occur is difficult, and I am unable to add any statistical
evidence upon the point, beyond that in two-thirds of
my cases the disease was attributed by the patient to
some such cause.   Gross states that in 11·7 per cent.,
or 23 out of 270 cases analysed by Winiwarter and
himself, the disease was attributed to injury, as blows
or contusions.

In the following case the cause and effect seem to
be closely connected :

Carcinoma of the breast in an old woman, aged 71,
*following a blow ; it was ulcerating, and on that account
was removed, and with good success.*—Susan C., aged 71,
came under my care in October, 1880, with an ulcerating carcinomatous tumour in her right breast, which
had appeared within a few days of a blow she had

received in the part ten months before admission. **Two** months ago this tumour had broken down and **discharged.** When seen a carcinomatous tumour occupied the upper half **of** the gland. The skin over it was infiltrated **and ulcerated.** The nipple was retracted. Axillary **glands** not to be felt. The breast was **removed,** as the ulcerating surface **was a source of distress** and weakness, and the **patient** did **well. The** temperature after the operation never rose above **99°.**

Eczema or ulceration of the nipple, as a precursor of cancer, is now a recognised cause, since it was pointed out by Paget, in the tenth volume of St. Bartholomew's Hospital Reports, published 1874, although in what proportion of cases the nipple trouble and the breast disease are associated is not yet evident. That eczema of **the** nipple may exist and get well, without giving rise to a cancer, is a clinical fact, **as** is likewise the formation **of** a cancerous **tumour** after the long existence of the nipple trouble.

Paget, in **his** paper, quoted fifteen examples, and described the disease as having "the appearance of a florid, intensely **red,** raw surface, very finely granular, as if nearly the whole thickness **of** the epidermis were **removed ;** like the surface **of** very acute diffused eczema, or like that of an acute balanitis. From such a surface there was always copious, clear, yellowish, viscid exudation. In some cases the eruption has presented the character of an ordinary chronic **eczema or psoriasis,** the eruption spreading beyond the **areola in** widening circles, or with scattered blotches **of** redness covering **nearly** the whole breast. The eruption **has resisted** all treatment, both local and general, and has continued even after the affected **part** of the skin has been involved in the cancerous **disease.**"

Butlin has added to this classical description of the **local** disease, as seen at the bed-side, some histological information, and has shown us how, as he believes, the carcinoma becomes a consecutive disease to the nipple affection.* "The activity and proliferation of the epithelium of the nipple is continued into the mouths of the galactophorous ducts; the same activity and proliferation can be traced along the ducts, in which the form of the cell is usually changed, and a spheroidal cell is substituted for the columnar cell, which naturally prevails there; these morbid conditions are continued deep down into the substance of the mammary gland, where the sacs become similarly affected."

"Coincident with the changes in the ducts and glandular epithelium, increased vascularity and the collection of cells resembling leucocytes are observed in the connective tissue immediately surrounding the glandular apparatus. But no direct connection can be traced between the changes within the ducts and sacs, and those without them, and the latter are subordinate in character and degree to the former. Not until the disease is advanced, and a definite tumour has been formed, which can be recognised with the naked eye as carcinoma, are cells resembling those within the ducts and sacs found in the surrounding tissues."

At a later stage of the disease the ducts and acini of the gland will be found filled with proliferating epithelium, which eventually escapes from the ducts by rupture or growth into the surrounding tissues, thereby producing the full formation of carcinoma.

Dr. Thin,† who made a minute examination of the skin of the nipple and breast tumours in four of these cases, two of them having been subjects of operation by Mr. Henry Morris, one by Mr. George Lawson, and the other by Mr. Manby, of East Rudham, has taken a different view as regards the nature of the

---

* Med.-Chir. Trans., vols. lx. lxiv.
† *British Medical Journal*, May, 1881.

skin affection. In two cases of the disease which he had an opportunity of examining before operation, the clinical evidence was, in his opinion, against eczema, and in favour of an infiltration in the superficial layer of the cutis. Microscopical examination he considers to be decisive on this point, proving that the **disease** is not an eczema, the destruction of **the cells of the** epidermis and the breaking down of **the** bundles of white fibrous tissue in the papillary layer of the cutis being in excess **of** the changes which are found in cases of eczema. In distinguishing clinically this malignant dermatitis from eczema, Dr. Thin considers that the chief points to be borne in mind are the well-defined margin in the former, and the evidence, when the tissue is grasped between the **fingers,** of infiltration in the papillary layer. **In two cases** which came under his notice there was clear evidence that the disease began in a position corresponding **to** the openings of the galactophorous ducts, **and remained** limited in that position for some time before extending over the surface of the nipple. He considers the infiltrated condition of the areola, which in **some of** its features is so suggestive of eczema, to be produced by the action on the connective tissue of a fluid which escapes from the mouths of the ducts, and which possesses the corrosive qualities of cancerous epithelium. He believes that the first morbid change **is really** a cancerous change in or near the mouth of **the ducts, and** that the so-called eczematous conditions are secondary to these changes. In all the four cases examined by **Dr. Thin** the cancerous tumours were **of** the variety appropriately termed *duct cancer,* **the** cancerous epithelium taking its origin not in the acini of the gland but in the epithelium of the ducts ; the *fibro-carcinoma cysticum mammæ* of Waldeyer.

    Dr. Munro * relates **three** cases bearing on this

* *Glasgow Medical Journal,* 1881.

disease. In regard to one of them he states that it is his firm belief that the primary sore was from the very commencement malignant disease of the epithelium at the mouths of the galactophorous ducts, and that the supposed induced or secondary disease in the breast stands in the same relation to the disease as the more distant glandular deposits.

Dr. Duhring* relates two cases which, he believes, show "that the disease is not an eczema, but that it is a peculiar disease with a malignant tendency." He remarks that in his two cases nothing of a malignant nature was suspected until after the lapse of five and ten years respectively, and that itching was in both cases insignificant until the affection had existed several years, in this respect the disease differing decidedly from eczema, where itching is one of the first signs noted. In regard to the clinical features, Dr. Duhring corroborates the description given by Dr. Thin. The circumscribed, sharply defined lesion, and the slightly elevated border, are symptoms which, he states, do not obtain in eczema. "The brilliant colour of the lesion is striking and is more marked than in eczema, the absence of the eczematous surface characterised by an appreciable discharge, or by vesicles, pustules, or puncta coming and going from time to time, and the absence of exacerbation so usual in eczema, may also be referred to." A point to which he also directs attention is the infiltration, "which is firm or even hard, but is not deep-seated. It is rather superficial. In eczema, on the other hand, it is soft."

This local disease, if eczema or otherwise, is therefore neither to be treated lightly nor disregarded, and the thought that it will eventually lead to, if it be not from the first, cancer, should be a prominent one in the surgeon's mind.

*American Journal of Medical Sciences, July, 1883.

In most of the cases recorded, the nipple disease had preceded the breast disease from two to five or six years.

Gross and Oldekop's tables show that this eczema of the nipple preceded the carcinoma in 5 or 1·35 per cent. of 370 cases. Mr. Henry Morris* records two out of 305 cases. I have recorded but three examples in my 600 consecutive cases. I quote these three with others that I have extracted from my note-book.

1. *Carcinoma of breast, preceded by serous discharge from, and eczema of the nipple; operation; convalescence.*—Martha C., aged 33, came under my care in January, 1880, with a tumour occupying the upper half of her left breast, which had been growing for one year, with infiltration of the skin over it : there was likewise a local patch of eczema involving the areola, and half the nipple, which is retracted. She had, for some weeks before the tumour was discovered, a discharge from the nipple, and some scabbing at its apex. The axillary glands are enlarged.

The breast, tumour, and glands were removed with a good result.

2. *Carcinoma of the right breast, following eczema of the nipple; excision of tumour, which was a typical duct cancer; cure.*—Jane G., aged 53, a healthy woman, the mother of several children, all of whom she suckled without difficulty, came under my care on January 28, 1887, with a tumour in her right breast.

It had appeared four years previously, in 1883, as a small lump, the size of a pea, on the outer half of her right breast. It gave no pain, and she disregarded it. In August, 1885, the nipple became sore, and she sought advice. Some application was applied to the nipple, and the eczema improved, but never got well. The lump in the breast during this time had much increased. When admitted under my care into Guy's

* *Lancet*, vol. ii. : 1879. Med.-Chir. Trans., vol. liii. p. 49.

Hospital there was a patch of eczema, about the size of a florin, occupying the site of the right nipple, which had become obliterated. Behind and to the outer side of the eczematous patch there was a hard nodular tumour, measuring three inches transversely and about two inches vertically, in the substance of the breast. The skin over it was not implicated, and the tumour moved with the breast freely over the deeper structures. It was not painful. An enlarged gland could be felt in the axilla.

On February 5th the tumour on the breast and its axillary glands were removed, and a rapid recovery took place.

Mr. Targett, the surgical registrar and pathologist, examined the tumour, and gave the following report : The microscopic examination showed the changes of chronic eczema in the nipple and areola. There was proliferation of the epithelium along some of the ducts, and especially in the terminal acini. In the latter situation there was much small cell infiltration around groups of acini, and in this manner they were converted into the alveoli of fully developed carcinoma. It is thus a good example of duct cancer following on chronic eczema of the breast.

3. *Eczema of nipple ; cancer of breast.*—Eliza S., aged 62, who had been married 35 years, and had 6 children, came to me on February 19th, 1865, with an infiltration of the upper lobe of her right breast, which had existed for one year. For two years previously she had had eczema of her nipple, and during this time the nipple projected ; for three years before the eczema appeared it had been retracted. In June, 1865, the skin over the breast was infiltrated with cancerous tubercles. I then lost sight of the patient.

This patient's mother died from cancer of the breast, and two of her brothers of internal cancer.

4. *Paget's disease of nipple, followed by carcinoma*

destroyed with cautery.—Mrs. B., aged 60, a married, childless woman, came to me March 26, 1877, with eczematous ulceration of the right nipple and areola, of four years' standing. The nipple had gone, and much of the areola. The surface was ulcerated, but **not** thickened. Total destruction of the whole surface **with a** cautery was carried out, and the parts healed. Some months later the hardness reappeared, and ulcerated, and in 1879 the right breast was excised, and convalescence followed ; but the disease soon returned, and invaded the opposite breast. The patient died in 1882.

5. *Carcinoma of breast following eczema of nipple.* —Miss S., aged 57, consulted me in May, 1875, for an eczema of her nipple which had existed for many months. When seen by **me**, there was marked infiltrating carcinoma of the breast ; retracted nipple, and a red, raw surface in the position of the nipple and its areola.

Operation advised, but refused.

6. *Carcinoma of breast following eczema of nipple.* —**Mrs.** S., aged 45, the mother of 5 children, came to **me** in October, 1877, with an eczema or rawness of one nipple of two years' standing, followed by swelling, hardness of the breast, with retracting nipple. This increased for nine months, and then ulcerated. When seen there was an open cancerous ulcer of the breast **and** nipple. Her mother is said to have died from cancer of the face.

Operation advised, but refused.

Mental anxiety as a cause of carcinoma must not be omitted, although to adduce proof of the truth of the proposition may be difficult ; to say that it is an antecedent of many cases of carcinoma is, however, **a** fact which the experience of most surgeons would support. The late Mr. Charles Moore was a firm believer in its influence.

On the influence of locality in cancer.—
It is impossible to deny the influence of nationality
and of geographical position upon the frequency of
cancer. For in some countries little of it is seen,
whereas, in others, it is met with frequently in
certain districts, and in others but rarely. In
Turkey, Greece, Syria, Persia, North Africa, and Ice-
land, it is very rare, whereas in parts of India and
China it is more common. In Europe its relative
frequency is very curious, and Hirsch tells us that it
varies from nearly 10 cases in 10,000 inhabitants, as
met with at Trondhjem in Norway, and in Lombardy in
Italy, to 2 in 10,000 in Sardinia. In America the
death rate is about 4·5, and in England much the
same, the rate apparently in our own land having
increased during the last few decades; that ending
1859 having been 3·3, and that ending 1876 4·6 per
10,000. In all these statistical figures it may be
fairly assumed that in women half the cases are
affections of the breast and uterus.

Lastly, as one of the causes of cancer, and consequently of cancer of the breast, is the geographical
distribution of the disease. Dr. Haviland having
demonstrated by means of a cancer chart collated from
the records of the Registrar-General, and read before
the Society of Arts in 1879, that there are certain
localities in England in which cancer predominates
largely over others, and he certainly has shown that
these cancer fields lie along the low lying alluvial beds
of river courses, and the geological strata of the tertiary
formation. "Cancer fields," he wrote, "are situated
along rivers that occasionally overflow their banks."

Hirsch, however, is no believer in this theory or
explanation, for he reminds us that in Norway, cancer
occurs mostly in the mountainous districts, and at
considerable elevations; also that in Mexico the high
table-land is more subject to cancer than the low places.

By way of summary as to the causes of carcinoma, the following conclusions seem just.

1. That there exists in those who manifest **the** disease, and probably in many others who never live to do so, **a** predisposition to its development under **any exciting** cause.

2. **That this** predisposition is, to a degree, **due to** heredity, although sufficient evidence exists **to** suggest that the predisposition may be strengthened, or even acquired, either by long residence **in** low lying districts in which rivers are prone to overflow their banks, **or** by the long continued depressing effects of mental influences.

3. That the disease is, **under all circumstances,** primarily a local one, originating in a gland or **tissue** that has been the seat either of a direct injury, **or** that has undergone some degenerative change, the effects either of age, obsolescence, or some antecedent spurious functional activity, inflammatory action, **or** persistent local irritation.

CHAPTER XIII.

THE CLINICAL FEATURES OF A SCIRRHOUS CARCINOMA.

WHEN a cancerous tumour has fully developed, and more particularly the first two varieties, it has features of so marked a character that its nature can hardly be misunderstood. When it is developing, its early features are ill defined and uncertain, consequently a positive diagnosis of its nature is difficult if not impossible; and yet it is at the very early stage of the disease that a diagnosis is most essential, since **it is**

then, if at all, that a cancer may be regarded as a
local and consequently as a curable affection.

It is to be remembered that in the majority of
cases of carcinoma of the breast, the disease attacks
the breasts of women about forty years of age when
the procreative organs are verging towards their natural
period of functional decline; and when the breast as
a gland is either obsolete, or is passing into the stage
of obsolescence. It is, however, found in younger, as
well as in older subjects; when in the younger it
usually appears as an acute or active affection, when
in the latter, as a more chronic one.

Early local symptoms of carcinoma.—In
the majority of cases when discovered it appears
either as an ill-defined thickening of one of the lobes
of the gland, or as a nodular swelling fixed closely to
the tissue of the breast. It is generally found out by
accident, as in washing, and the attention of the
patient is rarely drawn to the part by pain; should
pain be present it will generally show itself as an
occasional shoot of pain, although sometimes as a
sensation of heat.

With these early symptoms present, how is a diagnosis to be made? for what affections may the disease
be mistaken? A few words upon these questions
may be of value.

Should the disease appear as a mere thickening of
one of the lobes of the breast, this clinical fact is
one, which of itself should excite the apprehensions
of cancer; for the thickening can only be brought
about by one of two causes, either carcinomatous
infiltration, or chronic inflammatory infiltration;
and it must be admitted that too often to the touch
these two forms of infiltration yield the same sensations.
In cancer, however, in a large proportion of cases, the
hardness is of a stony, unyielding character, and the
examination, if not rough, is well tolerated; whilst in

the chronic inflammatory condition the induration is fleshy and somewhat yielding, the same amount of manipulation likewise evokes pain. In the chronic inflammatory trouble there will likewise probably be more pain on gentle manipulation than is usually met with in cancer, and after examination the patient will show some signs of excitement or irritability. In this latter trouble, also, more than one of the lobes of the gland will probably be found coarse and knotty, although but one may be tender, and possibly the opposite gland will be equally affected.

In cancer, in brief, the disease starts in one lobe of a gland, and as an indolent swelling it is rarely associated with any signs or symptoms of local inflammation, such as may be made out in the chronic lobular inflammation of the breast, which is prone to attack women at the climacteric period. Cancer in its early stage is known more by the absence of symptoms than by anything more definite, by a painless, or nearly painless local, stony infiltration of a part. This stage of cancer may be described as the "early infiltrating stage." When the disease first shows itself as a nodular swelling fixed closely to the tissue of the breast, the same thought of carcinoma should be excited, but at the same time the possibility of the lump being due to the presence of an involution cyst should not be forgotten. The rounder the lump, the greater the probability of its being cystic ; the more irregular and nodular the swelling, the greater the probability of its being cancerous. A hard, nodular swelling is more likely to be cystic than inflammatory, and it is at least as likely to be cystic as cancerous. For diagnosis an exploratory puncture with a needle is frequently demanded ; or what is better, an exploratory incision.

As the disease advances, and reaches the stage which, for descriptive or clinical purposes, may well be called the "mature infiltrating stage," the diagnostic points become more marked. The induration of the lobe primarily involved has **become more** evident, and it has probably **spread** from its early position in a single lobe, either to a neighbouring lobe or to the soft parts covering in the lobe, **and** in this way changes will have **been** brought about **in** the skin, which are all equally characteristic, and differ only in the degree in which the skin with its subcutaneous tissues is involved. I have been in the habit **of** describing these changes, **as** will have been observed by the reader in perusing some of the **preceding** chapters, under three headings, namely, "Dimpling," "Puckering," and "Infiltration" of the skin, and I believe that this division tends to clearness.

Dimpling of the skin means more or less depression or cupping of the skin covering the infiltrated lobe, brought about by the contraction of the suspensory ligaments of the breast (Plate I. Fig. 1), which pass from the normal capsule of the gland to the **skin** itself. Where only a few of the ligaments of the gland are involved, the dimpling **or** cupping of the skin will be limited; where the ligaments involved are numerous, the area of skin depression will be extensive. This condition is well seen in Plate III. Fig. 1. The skin at this stage can, however, be rolled over the growth beneath, or raised from it. This symptom is never present in any connective tissue tumour of the breast, nor in an inflammatory affection.

The **puckering** of skin is a later stage than that described, for in it the **skin is closely drawn** down to the growth **beneath, and has practically become a part of** it **by the** process **that** has been described as local infection. (*See* Plate III. Fig. 2.)

The suspensory ligaments have contracted to the

full, and the skin can no longer be rolled over or
raised from the growth. The depression of the
skin is now still more marked than it was in the
" dimpling " stage, and the surface of the skin presents
to the eye a peculiar pitted appearance, which Mr.
Banks has aptly described resembling the pig skin of
which a saddle is made. The skin, on manipulation,
may feel, to the touch, fairly healthy, or it may seem
slightly harder than normal. This condition being
suggestive of the next or most diagnostic indication of
carcinomatous disease (the stage of local "skin
infiltration ") in which the skin has become involved
by the extension of the disease, by local infection.

In the infiltrating stage the skin is not only
drawn down and puckered in the ways already described, but it is bound down to the growth beneath,
and to the finger feels firm and indurated ; the induration being brought about by the infiltration of the
skin itself with carcinomatous elements.

The tumour at this advanced stage of the disease
seems to be one with the skin ; neither in the gland nor
in the skin can the surgeon clearly define the boundary
of the growth. In fact, it has no definite boundary ;
for from its central lobular starting place the epithelial
elements have spread by infiltration outwards in all
directions, and drawn all surrounding tissues into
its sphere.

In some examples of disease at this stage of trouble
the mobility of the tumour will have considerably
lessened ; what could before have been moved freely
with the breast gland upon the extended pectoral
muscle, is now somewhat checked in its movements,
in some cases to a limited, in others to a marked,
degree ; the degree of immobility of the tumour and
breast gland upon the pectoral muscle below being
the exact measure of the amount of infiltration of
the deeper tissues, in the same way as the "dimpling,"

" puckering," and " infiltration " of the skin is an exact measure of the infiltration of the superficial.

To test the mobility of a breast tumour upon the pectoral muscle, it is essential that the muscle should be placed upon the stretch by raising the arm. With the muscle relaxed, a tumour which is perfectly fixed to it may appear movable, the tumour and relaxed muscle moving together. With the muscle extended this fallacy will be avoided.

Retraction of nipple in carcinoma of the breast.—There is no greater fallacy than the very general assumption that retraction of the nipple is an essential symptom of carcinoma of the breast ; and no greater error than that with this symptom the diagnosis of cancer is confirmed, and that without it the disease must be of another kind. The truth is that retraction of the nipple is only an accidental symptom of cancer of the breast, met with when the disease is placed near the nipple, but not otherwise ; that is, when the disease is so situated that the ducts of the gland are drawn upon by the infiltrating process, and the nipple, as a consequence, is pulled, as it were, towards the infiltrated lobe. When this retraction of the nipple exists associated with a chronic infiltrating tumour of the breast, it is, as a symptom, one of great value ; indeed, as great as the dimpling and puckering of the skin over the tumour has been shown to be (Plate III. Fig. 2). By itself the symptom is of no special value, since it may have been a congenital condition, or an acquired one from some antecedent inflammatory or other affection. Again, should the infiltrating lobule be placed at the periphery of the gland, where, by its contraction, the ducts of the gland, as they pass to the nipple, will be but little affected, there will be no retraction of the nipple. Indeed, the nipple may, under these circumstances, be very prominent. The symptom, therefore, is only of value when found in

combination with other symptoms; by itself it tells
nothing.   Again, in some cases the nipple may become
retracted during the early stage of the disease, and at
a later period become prominent; the disease at an
early period of its progress causing traction upon the
lactiferous ducts, and at a later period of its course,
by steady infiltration, so
thickening the gland be-
neath the nipple, or the
nipple itself, as mechanically to lift it out of its
umbilical bed and cause
it to project.   In a third
class of cases the nipple
may become strangulated
at its base by the con-
traction of the carcinoma,
and   consequently   œde-
matous, and possibly ul-
cerated.   At a still later
stage,   from   the   same
cause,   the   nipple   may
slough off.

Fig. 4.—Retraction of the Nipple
in case of Adeno-sarcoma of a
woman, aged 54.

At times a retracting
nipple is the first point
which draws the atten-
tion of the patient to her
breasts, and leads to the detection of an early carcinoma.

One of the best examples of retracted nipple I
have seen was in a woman past middle life, who had a
central tumour of her breast (Fig. 4).   The nipple
seemed telescopically to be inverted, and to form a depression in the centre of a circumferentially raised
fleshy growth.   The tumour, when removed, turned out
to be an adeno-sarcoma, developed beneath the base of
the nipple, in the centre of the breast gland.   By its
growth it had steadily made traction upon the ducts

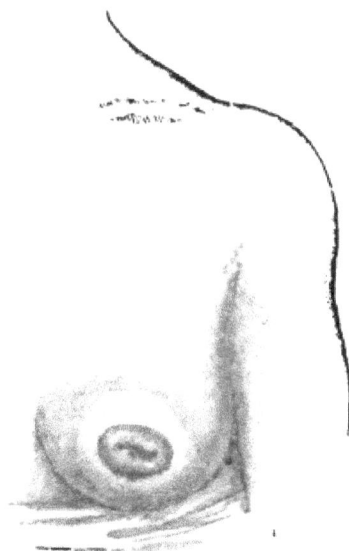

M—25

of the gland, and **as a** consequence drawn the nipple as described.   Simple tumours, however, rarely cause this retraction.   They may, by their **growth**, flatten a nipple out, or push it into some out-of-the-way place, but they seldom cause its retraction.   Where **this** symptom is present with a benign tumour, it **is due** either to the central position of the tumour, or to a congenital **or** acquired condition, such as **a former** abscess, **or** local inflammation.

Discharge from the nipple in cancer.— In the carcinoma fibrosum of the breast it is by no means common **for the patient to** complain of **any** discharge **from the nipple.   In** exceptional cases **this** symptom **may exist, and in** some cases before me **a** serous **or blood-stained** discharge had preceded the induration **of the breast** by some months.   In one **a** serous discharge was the earliest symptom ; **in** another a bloody discharge had existed for six **months** before any tumour was discovered ; in a **third case** a blood-stained clear discharge had existed **for** 2½ years before a **lump** was detected ; in the fourth case there had **been a** porter-like discharge from **the** nipple for four **years** before a tumour was **recognised.   In** all these cases, when they came **under** my care, the disease was clearly of **the true scirrhous** form, and unaccompanied with **any marked** cystic complication.

In a general way, **when** there **is** much nipple discharge **before** any **solid** neoplasm **is** discovered, the **disease has** been cystic in its origin, and carcinomatous **or** sarcomatous secondarily.   These cases will, however, receive **attention** in chapter xvi., which is devoted to cystic **disease.**

In the following cases the growth of **the** tumour was preceded or accompanied by **some** discharge **from** the nipple :

1. *Cancer of breast preceded by discharge from nipple.* —Jane E., aged 35, the mother of three children, the

youngest being eight months old, and all of whom she suckled, came under my care on October 26th, 1867, with induration of her left breast, dimpling of the skin over the breast, retracted nipple for six months, and discharge of blood from the nipple for five months. In January, 1868, the discharge from the nipple had ceased for two weeks, but returned copiously. On September 3rd, 1868, the breast was much larger and harder ; the nipple had retracted more ; the skin was puckered and infiltrated ; no axillary glands were enlarged. Operation was advised, but refused. On July 15th, 1868, the arm was œdematous, and she was sinking.

2. *Carcinoma of breast with imperfect nipple, and early discharge from nipple ; operation ; recovery.*— Sarah N., a married woman, aged 40, the mother of four children, all of whom she had suckled with her left breast, but not with her right, as the nipple was imperfect, came to me in August, 1872, with a general infiltration of her right breast, retracted nipple, puckering of the skin over the tumour, and enlarged lymphatic axillary glands. This disease had been coming on for four months, and attention was first drawn to it by the discharge of a curdy material from the nipple, which could be increased by pressure. The breast was removed and recovery followed. The tumour was of the scirrhous kind, with cystic elements.

3. *Carcinoma of breast ; bloody discharge from nipple ; operation ; well four years later.*—Mrs. G., aged 60, no children, for $2\frac{1}{4}$ years had blood-stained discharge from the nipple and some hardness of the left breast ; the discharge could be increased by pressure. January 28th, left mammary gland, nipple, and axillary lymphatic gland infiltrated with carcinoma. All were removed, and patient convalesced. The tumour was of the common scirrhous kind. Four years later this patient was well.

4. *Scirrhous carcinoma of the breast, associated with discharge of porter-like fluid from a retracted nipple, and followed by tuberculation of the skin and enlarged axillary glands ; excision* **of** *growth with glands ;* **return** *of disease in skin* **and** *opposite breast ; patient* **was** *alive four years after operation, or eight* **years** *after its first appearance.*—Mrs. W., a childless married woman, aged 46, came under my care in May, **1875**, with her left breast and skin covering it wholly infiltrated with carcinomatous material, and enlarged **axillary** lymphatic glands. **The** disease had been **slowly** coming for **four years, and** had commenced **as a glandular** induration **and some** retraction of **the** nipple, the nipple soon discharging freely some porter-like fluid. I excised the breast with the skin over it, and the axillary glands, and a good convalescence followed. Three years later some few tubercles appeared about the scar, and these slowly multiplied. In January, 1882, the opposite breast became involved, **but the patient's** health was good. I then lost sight of **her. The** disease had existed for eight years when I **last saw her,** and the breast had been removed **four years** previously.

"**Lymphatic infection.**"—As soon as the skin **has become** involved, or the tumour has become **fixed to the** deeper parts, another **stage** of **the disease has to be noted,** with probably another starting point **; for about** this time the glands associated with the tumour by their lymphatic ducts may be expected to become enlarged, and consequently infiltrated. Indeed, "lymphatic infection" may be expected to show itself as soon as it is clear that the disease **has** spread from **its** primary starting point, in whatever part of the breast that point may be, and invaded **the** connective tissue outside the gland, as evidenced **by** skin dimpling or infiltration ; this stage of the disease **according** with its entrance into the lymph spaces

of the glands, such being the true radicles of the
lymphatic vessels. When the glands are perceptible
to the surgeon's fingers, this lymphatic infection has
evidently advanced far, for in ordinary and not
very thin subjects there may be some enlargement
of the lymphatic glands of the axilla, or beneath the
pectoral muscle, without the fingers of the surgeon
being able to detect them. Surgeons are now familiar
with the fact that, in an operation, enlarged axillary
glands are frequently found where beforehand they
were not expected. In very thin people the diagnosis
of their presence or absence may with some accuracy
be determined. In fat subjects this cannot be said to
be the case. To make out the true condition of the
glands, they must be searched for by an incision
through the skin and fascia. To conclude that they
are not enlarged because they are not felt through the
integument would generally be an error.

The lymphatic glands in connection with a carcinomatous tumour may, however, occasionally enlarge,
and later on diminish and even assume their normal
dimensions, just as they are known to do as a result
of some inflammatory action of the part with which
they are connected. I have seen this occur in car-
cinomous breast cases on two very distinct occasions,
and do not doubt the fact ; and such a condition
is by no means uncommon in other regions, such,
for example, as the glands under the jaw in epithelioma of the tongue, or of the groin in penile
cancer. The explanation can only be that there was
associated with the carcinoma some temporary irrita-
tive cause, which gave rise to the lymphatic swelling,
which subsided with the cause.

In one case which made a lasting impression
upon me, the patient, aged thirty-four, the daughter
of a surgeon, had true scirrhus of the breast of
two years' standing, without evidence of lymphatic

complications. The lady was very thin, consequently her axilla was easily explored. I advised early operation in the case, and during the waiting period the glands suddenly became tender and markedly visible. Her father, under those circumstances, regarded the removal of the breast and glands unjustifiable, and so let things drift. In two months, under the influence of belladonna and glycerine as a local application, the glands steadily subsided, and when I saw her again they were not to be detected. I then excised the breast, laid bare the axilla to search for glands, which were not found, and the case did well. The patient survived the operation five years, and died from some internal disease, the nature of which was never made out by a post-mortem. It was believed to have been liver disease; probably it was carcinoma.

The second case was in a single lady, aged 36, from whom I removed a carcinomatous breast, and some enlarged axillary glands, in 1864. She recovered rapidly after the operation, and remained well for three or four months, when she was brought to me by her medical man for a swelling in the affected axilla, which had been detected about a week. The swelling was clearly glandular, nodular, and the size of a walnut. In three months, when I saw her again, the swelling had decreased under the use of belladonna and glycerine, and a few months later it had disappeared. The lady lived six or seven years longer, and died from some acute lung affection.

As the axillary or other glands enlarge, new symptoms and conditions show themselves; for the enlarged glands must be regarded as fresh centres of disease, which progress precisely in the same lines as the original affection. They increase in size locally, and infect neighbouring tissues by infiltration. The lymphatics that pass to and from the glands become

direct channels of communication of the disease, unless
they themselves become choked and their walls infiltrated, and thus cease to be tubes.   They feel, under
these circumstances, as solid cords, and may often **be**
traced in the skin as white tubes, thickened and nodulated (Plate **IV. Fig.** 1). As the glands enlarge, the veins
of the axilla, or above the clavicle, become pressed upon,
and the venous circulation of the arm of the affected
side, and of the side and shoulder, becomes interfered
with. Under these circumstances, a more or less rapidly
occurring œdema of the part involved is a new symptom.   The swelling may be either slow in its forma-
tion, or very rapid, this point turning upon **the**
suddenness and completeness of the venous obstruction, and upon **the** size and position of the vein **ob-**
structed.   Should the vein be one of small size, the
œdema will be limited ; should it be the axillary vein
itself the œdema will be general and rapid.   The **arm**
will enlarge and become congested, according to **the**
extent of vein pressure ; at times, from congestion,
it may be mottled or purple ; in other cases there may
even be local static gangrene.

    Very often, however, the œdema, which may
have set in suddenly, steadily subsides under treatment, by elevation, gentle pressure, and friction
directed upwards ; and when this takes place, it is
because the vein has either re-opened, or **the blood**
has found other channels through which it can return
to the heart.   This improvement is, however, rarely
lasting.

    With obstruction to the veins of the axilla, there
is generally pressure upon some of the nerve trunks
of the brachial plexus, and as a result pain is produced.
The seat of pain is determined by the nerve trunks
that are pressed upon ; in some of these cases it is
severe and most distressing ; too often it is of a
hopeless character, since there is but little prospect of

its being relieved by anything less than the death of
the sufferer.

Tuberculation of the skin (Plate IV. Fig. 4).
—If there **be** one symptom more typical of cancer of the
breast than another, it is the " shot-like " tubercular infiltration of the skin, which is met with in certain cases
of the disease. When this condition is present, there
can be no doubt as to the nature of the affection, for in
no other disease than carcinoma is the skin ever the
seat of deposits, which may be so small as only to be
perceptible to the finger when passed gently over its
surface, or so perceptible and palpable to the surgeon's
eye and hand as to show up as small wax drop-like
infiltrations of the skin over the deep tumour, as well
as over the surrounding parts. In some cases this
shot-like infiltration is very local, in others it is more
general. In some the skin appears to the eye mottled
irregularly red, and to the finger it is irregularly
indurated. In other cases the mottling and infiltration are general, and the skin feels universally indurated. Two or three tubercles in the skin associated
with **a** chronic infiltration of a lobe or lobules of a
breast, are, however, as typical of a local cancer as a
large number.

In exceptional cases these tubercles appear as
bloodless indurations of the skin, and then disappear
(*see* **case,** page 142) ; a withering atrophic change, apparently due to contraction of the fibrous elements
of the growth, **so chokes** or destroys what little
cell structure may **have** existed in its alveoli, as
to be followed by atrophy ; the atrophy in some
cases shows itself **by a** steady disappearance of
the tubercles, whilst in others the tubercles crumble
away and disappear. During this wasting process
there **is,** however, usually a developing one; and
whilst, **on** the one hand, tubercles disappear, others
show themselves, so that on the whole the disease

usually advances.   In a third class of cases, again,
the tubercular infiltration of the integument **may**
be an acute condition which shows itself in this form
from the first, or as a sudden outburst of development
accompanying a chronic carcinoma.   When these
tubercles appear in the skin, the surgeon may fairly
assume that they exist in other tissues surrounding
the local tumour, and their existence is to be accepted as evidence of local dissemination of the
disease.

**The brawny infiltration and tubercula-
tion of the skin.**—The tuberculated condition of
skin which has been just described is generally met
with in the course of development of the chronic
fibrous form of carcinoma or scirrhus ; but **the same
condition is** met with in that variety of **acute carci-**
noma which may well be described as **the brawny.**
It shows itself either **as a** rapidly disseminating
tubercular affection of the skin, or as an acute
œdematous or brawny infiltration (Plate IV. Fig. 3).
Under both circumstances the skin trouble is
secondary to the infiltration of the breast beneath,
although in certain cases the glandular and cutaneous
infiltration seem in point of time to coincide.

In the majority of these cases, however, the patient
will describe a lump as having existed in the breast
for some weeks, when suddenly the skin over it
became red and mottled.   If seen by the surgeon at
this time, the redness may be that of inflammation, for
which I have known it more than once mistaken, and
with the redness there may be local increase of heat,
a symptom which **helps** the error.   The surgeon must
not, however, be **misled by** these appearances and
conditions, for he **will**, on passing his finger over the
red and heated skin surface, at once make out that the
skin has lost its softness and smoothness, and feels
irregularly indurated ; he will, moreover, not be able to

raise the skin from the tumour, or, indeed, to separate
it from the tissues beneath ; the skin and tumour
appearing as one. At times the surgeon will be surprised to see interspersed upon this red and heated
surface, irregular spaces of apparently healthy skin, a
condition which by itself should be enough to suggest
that the symptoms are not due to inflammation, since
with inflammation no such spaces would exist. The
fact, however, that to the touch this reddened skin is
indurated and evidently infiltrated is sufficient evidence
to decide the diagnosis in favour of acute cancerous
infiltration.

One of the most typical cases of this kind that I
have seen was in a woman who was induced to put a
hot fig poultice to a tumour in her breast, which had
existed for some weeks. The poultice was applied
and kept in position for a day, and on its removal I
saw the woman. The skin of the breast was then,
over an area exactly corresponding to that which had
been covered by the hot fig poultice, red and apparently inflamed ; to the finger it felt brawny. I
suspected at the time that this condition was due to
acute cancerous infiltration of the part, and the progress of the case proved the suspicion to be true ;
for the redness never subsided, and in one week the
tuberculation of the skin was marked and characteristic. The case was clearly one of acute cancer, and
I believe the local hot poultice hastened its growth.

**The acute œdematous infiltration of the
skin and breast** is met with occasionally, and it
is without doubt the most acute and fatal form of
cancer found in the breast. It may attack the gland
much in the same way as the acute tuberculated form
which has just been described, but it may do so more
insidiously, or rather less actively, and appears more
as a rapid œdema of the breast and skin over it, with
or without retraction of the nipple, the whole gland

PLATE IV.

Fig. 1.

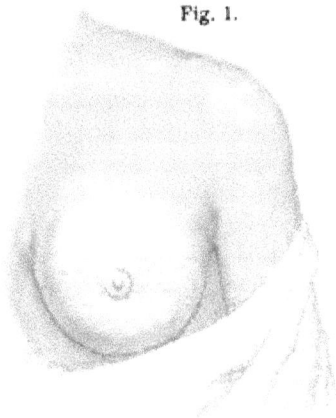

Fig. 2.

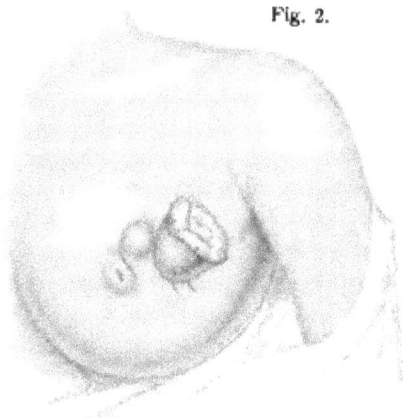

Fig. 3.

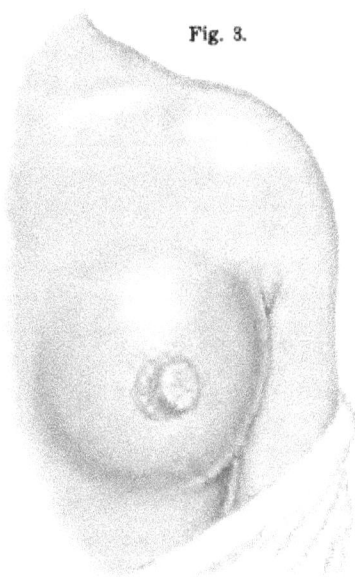

Fig. 4

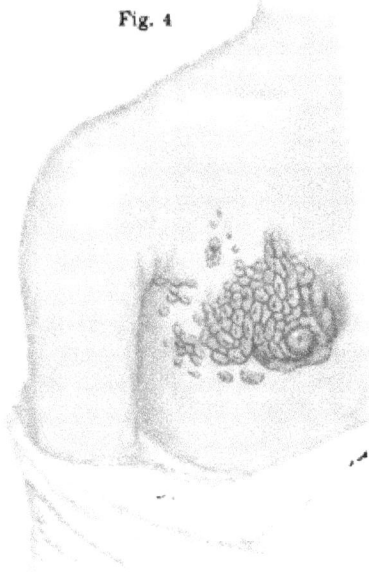

CARCINOMA OF THE BREAST.

1. Lymphatic Infiltration of Skin.   3. Brawny Carcinoma.
2. Open tuberous Carcinoma.   4. Tuberculated Infiltration of Skin.

and integuments covering it feeling to the hand infiltrated, and perhaps pitting on pressure. There may be no external redness or heat suggestive of inflammation, but only œdema, and this œdematous infiltration will probably be confined to the soft parts over the breast. In exceptional cases it will extend to a wider area. Such cases as these, like the last described, have been mistaken for inflammation, but a knowledge of the probability of this affection being cancerous should prevent such a mistake from being made.

With this brawny infiltration of the breast and skin over it, there is an allied condition which claims description, and I have been in the habit of describing it at the bedside as one of lymphatic absorption, in which the lymphatics of the skin of the breast appear as swollen white cords radiating from the nipple, with the intermediate skin slightly thickened from œdema, but not sufficiently so to pit (Plate IV. Fig. 1). These local symptoms are always associated with rapidly progressing disease, in which the original nidus not only spreads rapidly by infiltration, or "local infection," but also by "lymphatic infection," the lymphatics, as described, being apparently filled, if not choked, with the epithelial material which it is conveying to the lymphatic glands.

This form of carcinoma in its clinical features is not recognised as it should be. I quote, therefore, some few examples to illustrate its different points. It is to be observed that this variety of carcinoma attacks what appears to be healthy women ; indeed, many have a florid aspect of health. It spreads rapidly and kills quickly. Thus, in some of the cases quoted, the disease ran its course in five months. When complicated with pregnancy or lactation the disease is most active.

In the way of treatment little can be done. When

I have been tempted to operate I have often regretted doing so. In a general way it is best to leave the disease alone.

1. *Acute brawny carcinoma of the* breast in an *apparently healthy woman, commencing as an* erythema *with œdema three months previously ; death in less than five months from the first onset of the disease.*— Miss B., aged 47, a healthy looking woman, came to me in April, 1871, with a breast which was œdematous and brawny, and with the skin over it full of flattened **carcinomatous** tubercles. The disease had appeared three months previously as a **swelling** of the gland, and great redness with œdema of **the** skin, over the gland. The redness soon subsided, but the swelling remained and became harder. The disease spread rapidly to the parts around, and the patient died in June, 1871, exhausted from probably internal disease.

2. *Acute brawny carcinomatous* infiltration *of breast and skin ; death in five months from discovery of disease.*—Eliza B., aged 43, a healthy looking single woman, came under my care December 13th, 1869, with acute disease of her right breast, existing only two months. The whole gland was infiltrated, and **the skin over it** like brawn. The nipple was depressed, **and lost in** the surrounding elevation of the breast ; the axillary with the supraclavicular glands were enlarged. On February **17th** the arm was œdematous. On March 3rd the woman was sinking.

3. *Acute brawny carcinomatous infiltration of breast and skin over it ; death* in six *months.*—Ann R., aged 53, a married woman, the mother of eleven children, **the** youngest being 12, came under my **care** on June 1st, 1857, with an acute brawny infiltration of her right breast and skin over it, and œdema of the right **arm.** She had been perfectly well up to three months before, when she noticed a swelling in her breast,

which rapidly increased and became complicated with pain down the right arm, and swelling. When seen, the axillary and supraclavicular glands were much enlarged; the **breast** was like brawn, and the skin over it was evidently infiltrated with new elements, **and** œdematous. In one month all the symptoms were worse, and the skin became **the seat** of acute infiltration of cancerous tubercles. Cough also had appeared. On **September** 15th, that is within six months of the first appearance of her trouble, she died with internal cancer from exhaustion.

4. *Acute brawny cancer of both breasts, proving fatal in nine months.*—Ann W., aged 49, the mother of six children, all of which she suckled, came under my care **on** January 26th, 1865, with an enormously swollen, œdematous, and indurated left breast, of eight weeks' standing. There was pain in the part, which was worse at night, and there was at times increase of heat in the **gland.** **The axillary** glands were uninvolved, the integument over the breast was œdematous, **and pitted on** pressure. In the **course** of three months **the axillary glands** became enlarged. The œdematous integument over the breast became as hard as brawn, and ceased to pit; it often also became red as if **inflamed.** Three **months** later the arm became involved by extension of the brawny infiltration, and the right breast became similarly implicated, the disease in the right gland passing through all the same stages as it had done in the left; the right arm also became swollen. The patient lived **nine** months after the discovery of the **tumour, and died** from asthenia, with probably internal cancer.

5. *Brawny carcinoma of breast; excision of gland; rapid death of patient from internal disease fourteen months after discovery of the tumour.*—Eliza B., aged 32, **a by** no means unhealthy looking single woman, **came under my care** on January 7th, 1870, with a tumour

in her right breast of eleven months' standing. It had grown slowly up to a month ago, when it spread rapidly. At present the whole gland is generally infiltrated, although movable upon the pectoral muscle. The skin over it is like brawn ; the nipple is retracted ; the axillary glands are not to be felt. The tumour and skin over it were excised, and a good convalescence followed, but the woman died April, 1870, from internal cancer.

6. *Acute brawny carcinomatous infiltration, attacking breast damaged by suppuration ; death in fifteen months.*—Mary L., aged 36 (suckled five children, youngest nine months old), $3\frac{1}{2}$ years ago had abscess in her left breast, but has been able to suckle with it since. January 7th, 1869. Three days after her last confinement, nine months ago, she had abscess in the breast, which opened in many ways ; sinuses discharged for seven weeks, and then closed, after which a lump appeared in the breast, which has rapidly increased. At present the left gland is generally infiltrated, and the skin over it is like brawn, and full of tubercles ; axillary glands enlarged ; nipple drawn in. Six months later powers rapidly failing.

7. *Acute brawny carcinoma of breast in a pregnant woman, following abscess in the breast three years before.*—A woman, seven months pregnant, aged 36, came under my care in February, 1886, with an acute brawny infiltration of her left breast and tuberculation of the skin over it. It had commenced as a lump seven months ago, with her pregnancy, and has steadily increased. The lump had appeared at the seat of an abscess she had three years before.

8. *Acute brawny cancer of one breast, with lymphatic infection, and a local cancer in the other.*—Harriet F., aged 28, a married woman, with one child four years old, which she had suckled with the *left* breast alone, came under my care in

December, 1885, with an enlargement of her left
breast, which had been coming for ten months, after
a blow. The lump grew steadily, and was the seat
of a sharp pain. Four months after **its** appearance
the nipple began to retract. A month before she
came **to** see me she poulticed the left breast; and
after this the skin, which was white previously,
became **red**. Two weeks previously she observed **a**
lump in her *right* breast. When seen, the left
breast was twice its natural size; the skin over it
appeared to the eye red, and to be the seat of innumerable tubercles, varying in size from millet seeds
to a threepenny-piece. The whole gland was hard,
and the skin over it brawny and fixed to the parts
beneath; the nipple was retracted. The lymphatic
glands in the axilla and above the clavicle were **enlarged**. In the axillary lobe of the opposite **breast
there was a nodule** of carcinoma the size of a walnut
infiltrating one of the lobes. In the skin over the
sternum there were likewise tubercles. No surgical
relief could be suggested.

9. *Brawny cancer of both breasts; death in twenty
months from onset of disease.*—Mrs. A. D., aged 41,
the mother of five children born dead, consulted me
on November 7th, 1869, with both breasts generally
infiltrated with cancerous disease, and the skin over
them brawny and tuberculated; both glands were
fixed to the chest, and the axillary glands were
enlarged. The disease of the right gland had been
coming for fifteen, and that of the left **for seven months**.
The left breast was ulcerating, this process having
commenced in the nipple. The woman looked well;
within six months she died with lung trouble.

10. *Acute cancer of breast, with lymphatic infiltration of skin; excision; convalescence.*—Fanny A., aged
35, came under my care on October 27th, 1883, with a
tumour in her left breast. She had had one child

eleven years ago, and suckled, though with sore
nipples. Three months before admission she discovered a lump in the centre of her left breast,
the size of a bantam's egg; this had steadily increased. When seen, the whole breast seemed to be
generally enlarged; the nipple was retracted; and
the skin over the tumour showed white lines radiating
from the nipple ; these lines, which felt by the finger
like hard cords, were clearly lymphatics. The
skin was slightly œdematous ; no axillary glands felt.
On November 5th the breast was removed and the
axilla explored, some glands being found. The patient
did well. The tumour was clearly carcinomatous, but
of the softer kind. The skin was evidently infiltrated,
and the lymphatics in it choked with cell elements.
The external view of this condition is well seen in
Plate IV. Fig. 1. I am unable to trace the subsequent course of this case.

11. *Acute brawny carcinoma in a woman, aged* 71,
following a blow.—Charlotte M., aged 71, came under
my care, July, 1884, with an acute carcinomatous infiltration of her right breast, with lymphatic infection
of the breast and tuberculation of the skin over the
tumour. All these symptoms had appeared within
one month, and had followed the appearance of a
lump in the breast, which had made its appearance
five months after a severe blow upon the part.

12. *Brawny carcinoma of left breast in a woman
aged* 75.—Laura D., a widow, aged 75, came under my
care in December, 1872, with a hard, tuberous, carcinomatous growth in her left breast, which had commenced six months previously as a small lump the
size of a plum, and had steadily increased. When
seen, the tumour was the size of a fist, and occupied
the whole gland. It was very hard, and felt as if
made up of several distinct tumours. The skin over
the tumours was *œdematous,* and felt like brawn.

The nipple was retracted and the axillary glands were enlarged.

13. *Carcinoma in an ill-developed breast, which had been damaged by an abscess; lymphatics infiltrated, and skin tuberculated; operation; death within a year.* —Eliza B., aged 62, the mother of three children, all of whom she suckled with the right, but not with the left breast, as it **was** not fully developed. During the second lactation an abscess appeared in the left breast. One year before she consulted me, on November 23rd, 1879, she noticed a small lump in the left breast. This increased slowly up to four months ago, when it became painful. When seen, the whole left breast was infiltrated with some neoplasm; **the** skin over it was adherent; and radiating from the nipple, the lymphatics in the skin appeared as hard, white, radiating lines. There were also in the skin many pink tubercles. The nipple was retracted; the breast itself was nodular and hard; the axillary glands were apparently unenlarged. November 28: Breast removed and axilla explored, some enlarged glands being removed. The patient did well. Temperature after operation never ran above 100°. A section of the growth showed a perfect specimen of carcinoma of the softer variety, which had involved the skin generally. The white radiating lines were clearly lymphatics infiltrated with cancer elements. The patient did well after the operation, but died within twelve months from a return of the local disease **and** chest complications.

14. *Acute tubercular carcinoma affecting breast with depressed* **nipple;** *death in fourteen months.* —Ann G., **aged 46,** the mother of twelve children, the youngest being one year old (she never suckled with her right breast, as the nipple was retracted), came under my care January 2nd, 1868, **with acute** carcinomatous infiltration of her right

N—25

breast, with nodular skin tubercles over the gland, and enlargement of the axillary glands. The disease had commenced ten months previously. On May 8th, fourteen months after the discovery of the tumour, she sank.

15. *Acute tubercular carcinoma of both breasts and integument of chest and abdomen; ulceration of left breast.*—Harriet L., aged 45, a married, childless woman, came under my care in July, 1879, with a cancerous infiltration of both breasts, also of skin over the breasts, and covering the sternum, chest, and abdomen. The skin looked red, and felt hard, nodular, and fixed to the parts beneath. Both nipples were retracted, and the axillary glands on both sides enlarged. The disease had commenced two and a half years previously as a small hard lump behind the nipple of the left breast, which had appeared two months after a blow upon the part. Two months later the nipple retracted, and the tumour spread. Eight months ago the skin became discoloured and lost its softness, and about this time a lump was noticed on the opposite breast. Three months ago the skin over the left breast ulcerated, and there has since been discharge. Four months ago she had pain in both axillæ, and two months later œdema of the left arm. Her general health was good, and she had no ill aspect. Nothing could be done.

Ulceration of a cancerous tumour.—In an early page of this work, it was pointed out that cancerous tumours, like other neoplasms, are shorter lived than the normal tissues in which they are placed, and that they will to a certainty, in the course of their progress, undergo degenerative changes, if not death. Under such circumstances any carcinomatous tumour, or, indeed, any tumour of the breast, cancerous or otherwise, will sooner or later break down either in its centre or on its surface, and give rise in the latter case to a cancerous ulcer (Fig. 5), or in the former case to a cancerous cavity, which discharges from its centre after the manner of an abscess (Plate IV. Fig. 2).

The cancerous ulcer will present features which will correspond to the form of cancerous tumour from which it originates. Should it follow the degenera-
tion of the atrophy-
ing scirrhous variety described in class 1, the ulcer will be of the most indolent kind, and appear simply as a surface deprived of its upper layer of epithelial elements, and discharging hardly more than the crumbling *débris* of a withering tumour. This va-
riety of ulcer may,

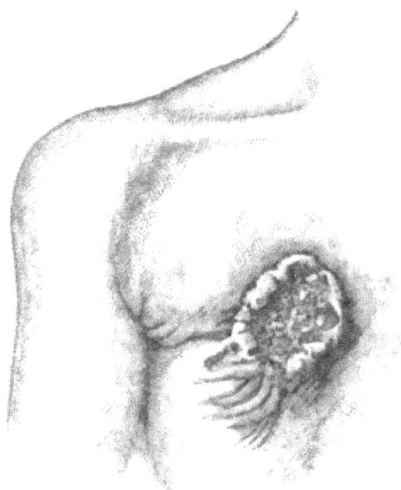

Fig. 5.—Carcinomatous Ulcer of Breast.

indeed, dry up, but it can hardly be described as healing (Plate VI. Figs. 1 and 2).

Should it be consecutive to the breaking down of the more common variety of carcinoma fibrosum (class 2), the ulcer will be typical of what is known as a cancerous ulcer (Fig. 5); that is, it will present an irregular surface, discharging serous fluid containing the *débris* of degenerating and breaking down epithelial tissue, with raised, thickened, infiltrated, and everted edges, which are so unlike the edges of any ulcer the result of inflammation, as to forbid an error in diagnosis being made. The process of degeneration which originated the cancerous ulcer steadily continues, and consequently the ulcerating process spreads; for a cancerous ulcer rarely heals; it is almost always a spreading one, and continues until the patient is exhausted. Cancerous ulcers do not fungate; fungating growths are usually sarcomatous.

When the degenerating or dying process commences in the centre of a cancerous growth (as it does when the disease is more of the tuberous than of the infiltrating variety of cancer, as described in class 3), the ulcerating skin process will be consecutive to the formation of what clinically might have simulated an abscess; since it would have commenced as a softening process of the centre of the dead and dying neoplasm, and continued as an extension of the process through all the parts which covered in the cavity, till the tissues outside gave way and permitted the escape of the cavity's contents.

The material that escapes from the cavity would be dead tissue mixed with the fluids of the dying parts; the walls of the cavity would be degenerating cancer; the edges of the opening through which the dead tissue escaped would likewise be cancerous, for they would appear as thick, raised, elevated, infiltrated, pouting, everted lips, from between which a thin, ichorous, and probably foetid discharge would escape, the discharge consisting of the *débris* of dead and dying tissue.

When the skin is tuberculated in any of the ways previously described, there may be ulceration, the tubercles to a certainty eventually undergoing degenerative changes and breaking down. Under these circumstances there may be many ulcers, but each will have its own special character, which will correspond to one or other of those which have been described. In the brawny form of cancerous infiltration the ulceration may be extensive, the breaking-down process generally commencing in the skin tubercles and spreading **to the deeper parts. This** process is rapidly exhausting, and **soon** terminates life.

In exceptional cases the whole or part **of a** tumour may die outright, **and then** slough off. This result happens **at times** when **the tumour** becomes **the seat of an** inflammation, and particularly of **erysipelas.**

By **way of** summary, **I quote a** paragraph **from** Gross's excellent work on mammary tumours, not that I can endorse from an analysis of my own cases his carefully worked-out and definite averages, for my own cases do not give me with sufficient accuracy the necessary data, but because upon the whole I believe his deductions to be practically valuable :

" From the preceding facts we learn that carcinoma evinces a remarkable disposition to infect the adjacent tissues, **and that** it progresses at first towards the **surface. The skin** is invaded in 68·9 **per cent., deep** attachments ensue in 21·9 per **cent., and** the opposite breast **suffers in** 3·6 per cent. of all instances. The occurrence **of** local dissemination is, moreover, indicated by the formation of circumscribed nodules in the skin in 10·6 per cent. ; in the subcutaneous connective tissue **in** 8·39 per cent. ; in both these situations, as in **the** cuirass form of cancer, in 1·95 per cent. ; in the pectoral muscles in 7 per cent. ; in the intercostal muscles in 1·35 per cent. ; **and,** finally, **in the** ribs in 2·8 per cent.

" In the order of the date of these appearances, we **may look for** extension to the superficial fasciæ and skin in 14 months, for ulceration in 20 months, for fixation to **the** chest in 22 months, and for invasion of the second breast **in 32 months."**

Clinical features of a medullary (tuberculous) carcinoma (Plate III. Fig. 4 ; Plate IV. Fig. 2 ; Plate V. Fig. 4).—In this form of carcinoma the cell element predominates largely

over the fibrous, and there is no tendency to such
contraction of the growth as is well known takes
place in the scirrhous variety. The tumour consequently grows more rapidly, assumes a more
lobulated outline, is softer to the touch, as well as
more elastic and semi-fluctuating than scirrhus.
In it also the nipple is flattened out, not retracted.
The tumour may grow to a great size in a few
months. The last example I saw was the size of a
cocoanut. As the tumour grows the skin becomes
stretched, and full blue veins become very conspicuous
on its surface ; later on the skin becomes adherent to
the tumour by infiltration ; it may even assume a
red and inflamed appearance, the redness being really
due to acute infiltration. To the finger the skin will
feel, as previously described, œdematous or brawny.
At a still later stage ulceration will occur.

Before this stage has been reached, indeed probably before the skin has become fixed to the neoplasm, the axillary or clavicular glands will have
enlarged ; when this lymphatic infection has taken
place the disease goes on rapidly. This soft form of
carcinoma, like the scirrhus, is an infiltrating affection ; it is rarely if ever encapsuled ; it differs from
the scirrhus form more in its mode of growth than
anything else. At times the cell growth increases
so rapidly that it forms a mass, which, being embedded in the mammary gland, has a circumscribed
appearance, suggestive of being encapsuled. In other
cases the disease may develop within a cyst, and so
really be encysted. In such cases the presence of
other cysts will explain its appearance. It has, however, doubtless often been mixed up with sarcoma,
and I am convinced that I, in common with others,
have fallen into this error. Cases which have passed
under my care during the last ten years or more,
which I should formerly have described as examples

PLATE V.

Fig. 1.

Fig. 2.

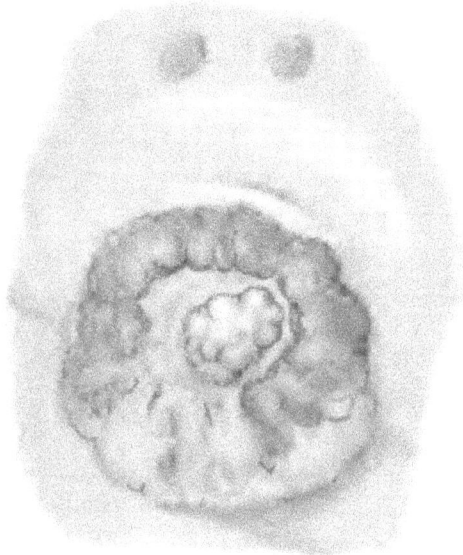

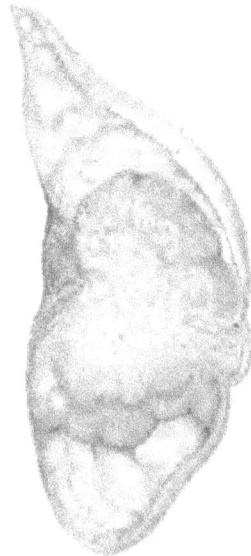

Fig. 3.

Fig. 4.

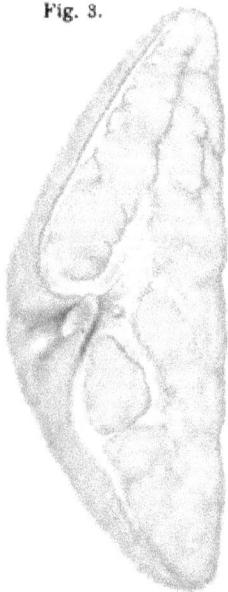

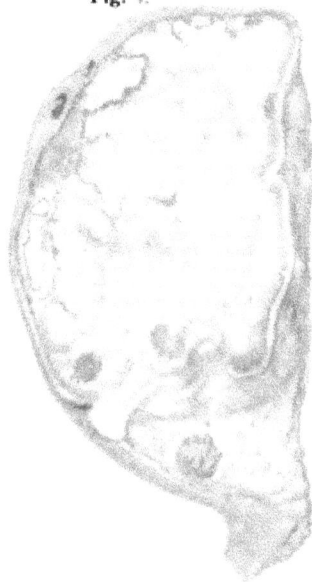

CARCINOMA OF THE BREAST.

1. Paget's Disease of the Nipple.   3. Section of Scirrhus.
2. Section of Carcinoma.   4. Section of soft Carcinoma.

of medullary carcinoma, I have called adeno-sarcomata; and I may say that I have not seen for many years, until quite recently, a good example of this affection. On looking into the literature of the subject, the above conclusion became very evident, and cases have been described of this affection which would have been better called acute carcinoma or rapidly growing sarcoma. For in carcinomatous, as in sarcomatous, tumours, there is great variety, and this variety is determined in both classes by the relative proportion of the fibrous and cell elements. When the fibrous element predominates, a scirrhous tumour is said to exist if the neoplasm be of the carcinomatous type; whereas if the cell element predominates, it will approach the character of what has been described as medullary carcinoma.

Under these circumstances it would probably be wise if the term encephaloid variety of carcinoma was abolished, for clinical experience tells us that in the long chain of varieties which forms the carcinomatous group of tumours, there may be innumerable degrees as to density and cell growth.

Clinical features of a colloid carcinoma. —This is probably always a degenerative change of a carcinomatous tumour, although in exceptional cases the colloid change is such an early feature of the disease as to suggest that it may be a primary growth.

It is found mostly in women after 40, as is carcinoma, but it may be met with in younger as well as in older subjects. It is rarely an acute affection; indeed, it seems to be of slower growth than the majority of cases of scirrhus. It begins as a lumpy swelling in a lobe of the breast, not as a simple induration as in scirrhus, which steadily increases. At first its growth may be slow; later on it becomes rapid. The outline of the tumour is usually bossy. The disease is generally preceded by nipple discharge, and this discharge is usually sanious (Plate VI. Figs. 3 and 4).

There is at times retraction of the nipple, but this symptom is not constant. **To the** fingers it feels firm, but elastic, and the bosses upon the tumour **are** suggestive of cysts. The discharge from the nipple supports this view. In the course of its progress the skin will become adherent, and later on ulcerated ; through the ulcerated openings the gelatinous or colloid fluid will ooze. The edges of the ulcerated opening are not infiltrated as in carcinoma, but more resemble **the** sharply cut edges of the cystic sarcoma**tous** growth.

Cases have been recorded of colloid carcinoma in which the skin has become tuberculated and the tumour fixed to the pectoral muscle. I have not **seen** those symptoms associated with it.

In the fifth volume of the Pathological Society's Transactions, a case of the late Sir **W**. Fergusson **is** recorded, in which recurrence after removal took place on at least two occasions. The patient was originally **under** Mr. **B. Travers's** care in 1845, when the breast was removed. Three years later a tumour was taken away from the cicatrix ; and five years later again Sir W. Fergusson excised several tubercles from the same part. The disease **had** therefore existed **for** eight years, and then seemed to be local. The patient was at last operation aged 38.

The disease may spread by lymphatic infection, though this rarely takes place till **a** late period of the case ; indeed, the disease may have existed for some three, six, or more years before this symptom appears. In a case of my own the disease had existed for seven years, and it was then quite local.

In brief, colloid carcinoma **is a** slow-growing tumour, and the least malignant variety of carcinoma ; it approaches, in fact, more the benign than

PLATE VI.

Fig. 1.

Fig. 2.

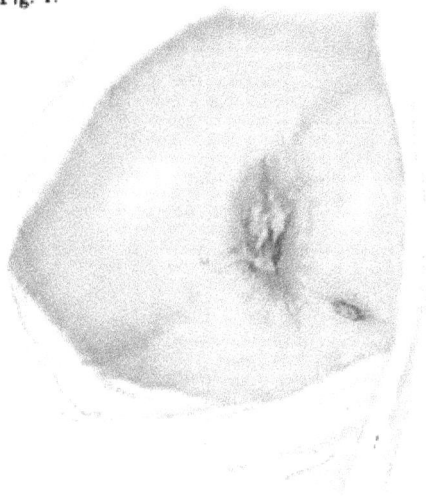

Fig. 4.

Fig. 3.

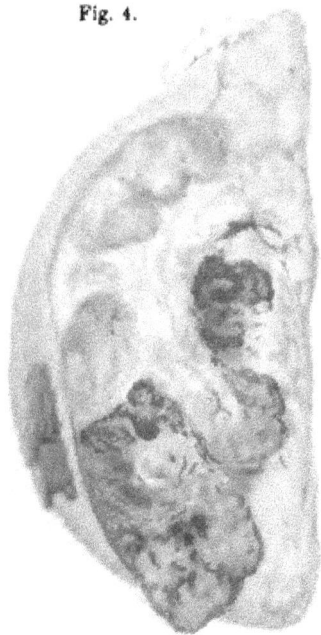

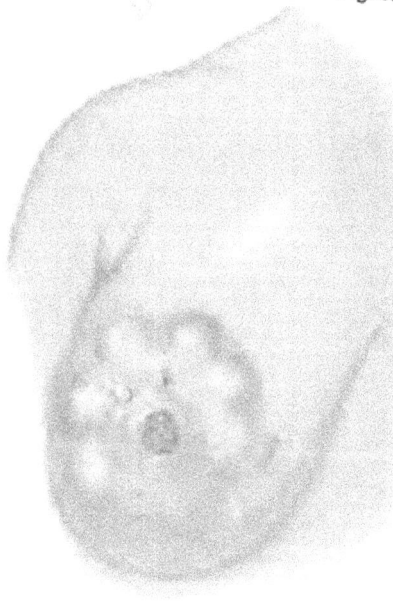

CARCINOMA OF THE BREAST.

1. Carcinomatous Ulcer.   3. Colloid Carcinoma.
2. Section of same.   4. Section of same.

the malignant growth. It is, however, difficult **to** diagnose before the degenerating or ulcerating stage has been reached, when the discharge of the glutinous, coloured, white-of-egg-like material from the ruptured cavity, and the well-defined uninfiltrated edge of the open wound clearly reveal its true nature. **In its** early state it never has the hardness of a scirrhous carcinoma, nor the roundness of an ordinary sarcoma, as seen by the drawing (Plate VI. Fig. 3); it has a bossy outline, more like an adeno-sarcoma or cystic sarcoma than anything else, and for the latter it is the most likely to be mistaken. I have seen but two examples of this affection, the one I now give an abstract of, and a second very similar to it in a woman, aged 56, who refused operation, and of whose future history I cannot learn anything.

Colloid disease of the breast of seven years' growth ; removal of gland ; seven years later carcinoma of the nipple of the opposite breast, following a local eczema ; no return of colloid disease.—Charlotte J., aged 65, **a** healthy, married, childless woman, came under my care on April 2nd, 1873, with a tumour in her right breast. The disease was first observed seven years before as two small lumps close to the nipple ; a year later these lumps, which had not increased much in size, had changed colour. Since then the tumour has slowly and painlessly increased. **Six** months ago, when the tumour began **to grow rapidly, pain** appeared ; and three months later the tumour **burst** and discharged a thick blood-stained fluid, **which has** continued ever since. On admission the tumour has a livid appearance, and measures six inches in diameter (Plate VI. Fig. 3). It is clearly lobulated, and each lobule fluctuates. The opening **into** one of the lobules is clean cut, and from it flows a thick glairy jelly-like coloured fluid. On April 8th the tumour was excised, and a good recovery followed.

On making a section of the growth its lobular nature was very marked. Each lobule was made up of a coloured jelly-like material, this caseating in parts. Under the microscope, in the fresh state, it showed a delicate stroma of connective tissue forming loculi, in which large granulation-like cells were found (Plate VI. Fig. 4). The patient remained well for more than seven years, when so-called eczema attacked the nipple with its areola of the opposite breast, and as this did not heal after six months' treatment she came to me. I then found, May 4th, 1881, a typical example of Paget's disease of the nipple, the base of which was indurated, and the nipple retracted. The lymphatic glands were not enlarged. The cicatrix of the operation upon the opposite breast was quite healthy. The woman at this time was 73, so no operation was suggested.

I am able to report that this woman was alive and in good health in 1886 ; that the local carcinoma had but slightly increased ; and that there was no return of her colloid disease.

In this case, therefore, a colloid carcinoma, which had existed for seven years before it was removed, showed no signs of return thirteen years later. It will be interesting to observe whether the carcinoma of her left breast, from which she is now suffering, will undergo a colloid transformation.

CHAPTER XIV.

TREATMENT OF CARCINOMA OF THE BREAST.

IF the view of carcinoma taken in the previous pages be correct, and the disease is to be regarded primarily as a local one, there can be no question as to the

principle of practice which should be followed in its
treatment, and that that principle can be summed up
in the one word, removal ; meaning by that word
the entire removal and complete extirpation of the
local affection. This practice is based upon the
conviction that **if** the smallest particle of diseased tissue be left behind, what remains will continue to grow ; and that the partial removal of **a**
carcinomatous growth tends to induce injurious
changes in, as well as a more rapid growth of, what
is left, whether that remnant be left in the seat of the
original growth, or **in** the lymphatics which pass
from it.

Charles Moore, of the Middlesex Hospital, **saw**
this so long ago as 1867, and in a valuable paper,
" On the Influence of Inadequate Operations upon **the**
Theory of Cancer," * laid it down as a **law that "the**
one important point, both for practice and theory,
is to remove the whole. The least remnant is capable
of growth, and may spring up into a new tumour
with all the energy of the first." " When any texture
adjoining the breast is involved in or even approached
by the disease, that texture should be removed with
the breast. This observation relates especially to the
skin, to lymphatics, to much fat, and to the pectoral
muscle. The attempt to save skin which is in any
degree unsound is of all errors, perhaps, the most
pernicious, and whenever its **condition is** doubtful
that texture should be taken away. A broad scar,
and the stretching and compression due to its subsequent contraction, appear **to be** especially satisfactory. In the performance of the operation it is
desirable to avoid, not only cutting into the tumour,
but also seeing it. No **actual** morbid texture should
be exposed, lest the active microscopic elements in it
be set free and lodge in the wound. Diseased axillary

* Med.-Chir. Trans., vol. l. p. 245.

glands should be taken away by the same dissection as the breast itself, without dividing the intervening lymphatics, and the practice of first roughly excising the central mass of the breast, and afterwards removing successive portions which may **be of** doubtful soundness, should be abandoned." These remarks of Mr. Moore's had reference rather to the practice of the day, in which the tumours were removed from the breast, and the nipple with portions of the gland left, rather than to the general exploration of the axilla for enlarged glands as recently advocated.

Thiersch also laid down the maxim that all operations upon epithelial **cancers** should **be** conducted **solely** with reference to their thorough eradication ; that all epithelial cancers should be operated upon if they can be effectually ; that a partial operation is, as a rule, a surgical blunder. In more recent times this principle of complete removal of **the local** disease has been considerably extended, and we find Mr. M. Banks, of Liverpool,* Mr. Pearce Gould, of London,† with Küster, of Berlin, and others, advocating **in** no feeble way the routine practice of exploring the axilla of every patient from whom a cancerous breast **is to** be excised. Küster in support of the practice **asserting** that at least one-third of his cases treated **in this** way have remained well for three **years, the** period which leads Gross to regard **as equivalent to a cure.**

With the leading thought of complete removal, therefore, engraved on the surgeon's mind, the corollary is clear **that the** earlier the local disease is taken away the more favourable **the result** is likely to be ; for theoretically **it can hardly** be doubted that if, on the very first onset of the trouble, the disease could be extirpated, a

* *Brit. Med. Journ.*, Dec. 9th, 1882.
† Med. Soc. Proceedings, 1884.

large number of cures would have to be recorded.
If an epithelioma of the skin can be cured by an
early removal, surely it is a reasonable hope that a
carcinoma (epithelioma) of the breast will likewise be
curable if treated in the same way, that is, by an early
excision.

Early operation, as well as complete removal, are
consequently the two essential points to observe in
the treatment of this serious malady, if operative
interference is undertaken under any idea of bringing
about a cure.

On early removal.—When a carcinoma of the
breast has so grown as to present features which make
the diagnosis of the disease certain, the trouble has
far advanced; it has probably reached the stage
already described as that of mature infiltration; for
infiltration of the **gland** of a typical character, with
dimpling or puckering of the skin over the gland,
and possibly some retraction or dragging of the
nipple, are advanced symptoms of breast cancer, and
indicate to a certainty that the infiltration of the lobe
of the affected gland, which was primarily local, has
spread beyond its original sphere, so as to affect the
coverings of the gland, and more of the gland than
the single lobe in which it originated.

A breast thus affected should doubtless, as a rule,
be at once wholly removed; but it cannot be denied
that the removal of the gland under these circum-
stances is less likely to be followed by a complete
cure, that is, a permanent absence of the disease, than
might reasonably be looked for if the breast had been
taken away before these well-marked symptoms had
put in an appearance.

The removal of the diseased gland under the circumstances described, is, however, far more likely
to be followed by a good result, than could be anticipated if lymphatic infection complicated the case, such

infection being unquestionable evidence of secondary cancerous extension.

What local symptoms, **it** may therefore be asked, should be considered of sufficient weight to lead the surgeon to suspect a local cancer? and under **what** circumstances may this suspicion of a local cancer justify the surgeon in recommending surgical interference? A satisfactory answer to these questions will be found in the chapter devoted to the consideration of the clinical features of a scirrhous carcinoma (chapter **xiii.**); **but it** may confidently be asserted that it is by **the** removal of the disease in its early infiltrating state that the best hopes of the surgeon may reasonably **rest.**

It is not, however, to be supposed that such an operation as excision of the breast gland is to be recommended or carried out, when possibly the clinical symptoms are no more than a local infiltration of a single lobe, and that infiltration has not been proved to be cancerous, for such advice would be wrong; but **I would** strongly urge upon all surgeons the propriety **of** making an exploratory incision into any tumour of the breast which seems to be infiltrating, and which **may** probably prove a carcinoma, in order that the complete extirpation of the diseased breast may **follow,** should the incision into the diseased part clear up the doubts which hung **about the** case, and make the diagnosis of carcinoma certain; whilst, on the other hand, if the tumour of the breast, into which an incision has been made, turn out to be of a simple character, the means adopted for diagnostic purposes may probably suffice for its enucleation or **cure.**

The diagnosis of early tumour is uncertain, **writes** Marcus Beck (Heath's Surgery), and yet if the disease is malignant, the only hope of cure lies in early recognition of its nature. In these cases, therefore, an exploratory incision should always be made. If the tumour is one of the simple growths, its smooth surface will be seen as soon as the capsule is opened, and the mass can be turned out without difficulty. If it is a sarcoma, its soft structure and less perfect outline will be recognised. If it is a cyst or an abscess, the contents escape ; and if it is an indurated lobule, its leathery toughness, and perhaps the small retention cysts, will be recognised. If it is a scirrhus, the " creaking of the knife as it enters it, and the sensation like cutting an unripe pear, and its sharp hard edge, not separable from the surrounding tissues, clearly show its nature."

" If the case is a very doubtful one, a freezing microtome and a microscope should be at hand, and a section should be mounted from a small slice taken from the growth. In this way all doubts are set at rest. ' Waiting till symptoms develop,' in the case of cancer, means waiting till all hope of cure is gone."

All chronic indurations of the breast in women over forty years of age, not enfeebled from age or infirmity, in which there is absence of evidence of its being inflammatory, should consequently be explored by an incision with the view of the exploratory measure being followed by excision of the breast, should the induration prove to be a carcinomatous infiltration.

Exploratory operations under these circumstances I hold to be of great practical and scientific value, and it would be well if they were always made. It would be **far** better for a surgeon to make at times an exploratory incision into a breast tumour under an erroneous impression, than to allow a carcinoma in the stage of early infiltration to advance and involve other structures, when we know that it is at this stage of early infiltration that its complete extirpation is most likely **to** prove permanently successful.

When enlargement of the lymphatic glands complicates the case of carcinoma of the breast, the disease may still be taken away if the surgeon sees a reasonable possibility of the removal of the affected glands with the primary growth and the general condition of the patient justifies the attempt. To take away the breast gland and leave the affected lymphatic glands is but a partial measure, since the disease by the operation is only partially removed, and the lymphatic glands which are left behind are likely, after the removal of the breast, to increase rapidly, and become fresh centres from which new troubles may emanate.

When, therefore, it is expedient to take away any well-marked carcinoma of the breast, it is a wise practice, in the majority of cases, to lay open the axilla at the same time, and look for enlarged lymphatic glands ; since it is now the common experience of surgeons to detect them after an incision, where before none were suspected. I have adopted this practice very generally for about ten years, and have reason to be fairly satisfied with the results. But I have not done so as a blind routine. In thin women in whom an axilla can be well explored digitally, I have not made this incision ; nor have I, in certain other cases, particularly in feeble women or in those advanced in life, in whom I felt the extra measure, by its increased severity, might prove detrimental, and more particularly when the axillary manipulation has been efficiently made with the patient under the influence of an anæsthetic, and no enlarged glands were to be felt. The rule of practice ought, however, to be axillary exploration in the majority of cases of excision of the mamma for cancer ; the omission of the practice the exception.

When a carcinomatous breast has broken down and is ulcerating, it may be removed if not too firmly fixed to the deeper structures, the disintegrating change in the tumour affording, indeed,

an extra argument in favour of operation ; and should
the primary growth be associated with lymphatic
glandular enlargement, and there is no reason why
these infiltrated glands should not be taken away on
account of their great fixedness or the age or general
enfeebled condition of the patient, the surgeon should
take them away with the primary growth, **and so**
work **for** a cure.   But should this complete operation
be contra-indicated on account of the size of the tumour
relatively to the age or infirmity of the patient (the
removal of the breast with the lymphatic glands
being, without doubt, a graver measure than the
removal of the breast alone), the incomplete operation
or the removal of the breast alone may be carried
out ; this partial measure being undertaken solely as
one of expediency, and with the view of removing a
local source of distress and weakness, and of making
what remains of life more comfortable.

Should the tumour be firmly fixed to the pectoral
muscle or deeper parts, and the glands in the axilla
be found immovable, or those above the clavicle
enlarged, the surgeon had better leave the case
alone.

Should the tumour also be very large, and the
shock of its removal likely to be severe when the
age of the patient or her feebleness is taken into
account, non-interference should be the rule of prac-
tice ; whereas should the tumour be small and movable,
and the woman yet old, it may be right and expedient
to take it **away.**   An operative measure, which in the
middle-aged **or** comparatively healthy woman would
be justifiable, may clearly be the reverse in the aged
or enfeebled ; and a small tumour may often be taken
from a healthy old patient, when a large one in an
enfeebled middle-aged **woman** had better be left
alone.

Certain old women, particularly the thin and **wiry,**

o—25

bear operations very well, and such **may** frequently
be submitted to surgical **treatment which** their fatter
or feebler sisters would be unable **to support.**

**When carcinoma of the breast is asso-
ciated with pregnancy,** the treatment of the
case is, indeed, difficult.   To **let the** disease **of the**
mamma **go on** with the physiological process **of** pregnancy, **even up** to if not through lactation, is to give
the disease the best chance of making rapid progress,
and to advance beyond surgical interference, and yet
to operate during pregnancy **is a measure of danger.**

The certainty **of the evil of delay is,** however,
greater **than the certainty of the evil of** action, **and I
am disposed to think that upon the** whole the advantages **of action** are the greater, and the practice based
upon it should be advocated.   In the three following
cases in which this practice was followed, **the success**
was sufficient to justify the means.

This practice **is,** however, to be **followed with**
extreme caution, **and** after the fullest consideration of
the facts of the *individual* case.

1. *Carcinoma of breast and pregnancy of seven
months; excision during pregnancy ; natural labour ;
no return two years later.*—Mrs. S., aged 46, the mother
of many children, came to me on May 14, 1879, when
she was pregnant seven months, on account **of an acute**
carcinoma of her right breast, which came on with **and**
developed during pregnancy.   I excised the breast
and the patient did well. **She was** confined on July 4
and had a natural labour.   There **was** no return of
her disease two years later.

2. *Carcinoma* **mammæ ;** *pregnant* **seven months ;**
*operation during pregnancy; recovery ; survived opera-
tion two years.*—Mrs. S., aged **42,** the mother of four
children, the youngest being one and a half years, who
had **not** nursed with her right breast from defect of the
nipple, came **to me** on May 10, 1879, when pregnant

seven months, with her right breast infiltrated and
nipple ulcerated ; no enlargement of axillary glands.
The breast was removed and a good recovery followed ;
she was confined at the full period naturally, and had
no breast trouble.   She died December 3, 1881, or
two and a half years after the operation, from internal cancer.

3. *Carcinoma mammæ ; pregnant six months ; operation during pregnancy ; convalescence ; confined at
full time ; survived operation one year.*—Mrs. B., aged
48, the mother of eight children, the youngest being
seven years of age ; the first two of which she
suckled without trouble, and with the later children
used her right breast more than the left, consulted
me on February 28, 1876, when six months pregnant,
with an infiltrating tumour in her right breast of six
months' standing.   The whole gland was involved.   I
excised the breast, and a good recovery followed.   She
was confined at the full period and the labour was
natural.   She survived the operation one year.

When the question of operative interference is
raised in cases of carcinoma of the breast, the nature
of the carcinoma, the size of the tumour, and the
clinical history of every case should be considered.
A measure which would, under one set of circumstances, be right, would under another be wrong ; and
an operation which, in one variety of the disease,
would be expedient, in another would be quite the
reverse.   I propose, therefore, now to consider the
question of operation in the different forms of carcinoma to which attention has been drawn.

**Operations in the atrophic or withering
variety of carcinoma** are not to be advocated.
Such cases are very slow in their progress, and
although slow they are, as a rule, sure.   Women
with this variety of carcinoma almost always die
eventually from internal disease ; and it seems that

interference with the local disease often helps to cause
the rapid growth and increase of secondary deposits,
which at the time were not manifest.   It is probably
true that if the disease were removed at its first stage,
that of early infiltration, good results might be obtained, in the same way as it is hoped that by early
operation in the ordinary carcinoma fibrosum, like
good results would have to be recorded. But, as a
rule, patients suffering from this atrophic variety of
carcinoma do not seek advice till the disease has
existed for years, when in all probability it has
ceased to be a local affection, but has become disseminated in some way, and probably by means of
tubercles.   Under these circumstances, local interference, or the partial removal of the disease, is followed by what may generally be looked for, the
rapid development, or rather growth, of whatever
secondary deposits may exist, either as lymphatic
enlargements, or secondary metastatic, or disseminated
tumours.

These are, moreover, the cases of carcinoma which
live the longest. They are those that, when mixed
in statistics, tend to raise the average duration of life
to a higher level, and that induce surgeons to believe
that non-interference in carcinoma of the breast
generally gives a longer life. (*See* cases reported,
page 142.)

**Operative interference in the acute or
brawny variety of carcinoma** is not to be
advised. When I have been tempted to remove a
breast under these circumstances, I have rarely had
any satisfactory result ; a speedy return, or rather
rapid growth, of the disease having been the rule, and
a long immunity quite an exception. The truth is,
in this variety of cancer it is impossible for a surgeon
to know the limit of the local affection, if, indeed, it
be local ; and under these circumstances, when an

operation is made, it must be regarded as only a partial or incomplete measure.

The brief reports of cases of this variety of carcinoma, printed at page 188, will show how acute this trouble is, how rapidly **it becomes** diffused, and how futile operative interference is likely to be.

Treatment of carcinoma of both breasts. —There is no reason why both breasts, **when simul**taneously affected with carcinoma, should not be taken away, if by doing so the disease can be wholly removed ; and there is no reason why the second breast, when so affected, should not be amputated when it **has** become the seat of carcinoma after the lapse of weeks, months, or years following the removal of its fellow for a like trouble.

The only bar to a double operation at the same time, or with an interval of time, is either the presence of lymphatic, glandular, or secondary deposits, which cannot be taken away ; or a feeble general condition of the patient which would be unable to stand the depressing influence of a large operation.

I append some cases of this comparatively rare condition. In several, both breasts were removed at different periods ; in others, one only was operated upon ; a second operation, for various reasons, not having been performed ; in a third class no operation was undertaken.

1. *Carcinoma of both breasts; operations; recovery; the left was affected two and a half years after the right had been removed.*—Harriet W., aged 60, came under my care in March, 1884, with a typical carcinoma fibrosum of her right breast, which had been coming on as a tumour for eight months, although the breast for a year previously had been the seat of pain. There were no enlarged lymphatic glands. Two and a half years before she had had her left breast removed for cancer, which had been growing for several

years. When removed, it was an open ulcer. The axilla at the time of operation was not explored. On March 25th the breast was alone removed, and a good recovery followed. This woman was alive eighteen months later.

2. *Carcinoma of both breasts; operations at fifteen months' interval; death in five years.* — Catherine T., aged 50, a widow, came to me on October 22nd, 1866, with an infiltrating cancer of her left breast, a retracted nipple and dimpling of skin. It had been coming on six months. In December, 1865, I had removed her *right* breast for infiltrating cancer, and a good recovery ensued. The axilla was not explored. In March, 1867, the left breast was removed, and a good recovery ensued; no axillary glands were felt enlarged. November, 1867, tubercles appeared in the skin over the thorax. March, 1869, cicatrix of right side sound, of left contains tubercles; left axillary glands likewise enlarged. This patient died in 1870 from internal cancer.

3. *Carcinoma of both breasts; operations.*—Miss V., aged 34, consulted me in November, 1877, for carcinoma of the *left* breast of two months' standing. I excised the gland, leaving the axilla unexplored, and convalescence followed. She left, a few months later, for America, where, one year later, the disease returned in the *right* breast, which was likewise removed. She lived about two years after the second operation, and died from chest complication.

4. *Carcinoma of both breasts; one breast removed; survived operation three and a half years.*—Miss B., aged 55, who had always had a retracted nipple in her left breast, consulted me in October, 1868, for a tumour which had been discovered one week, and was attended with pain which had existed for six months previously. When seen, the left breast was generally infiltrated, although the skin over the tumour and the

lymphatic axillary glands were uninvolved. Breast removed ; axilla not explored ; good convalescence. October, 1869, tubercle in scar. February, 1870, tubercles multiplying ; *right* breast infiltrated very slightly. Died March, 1872, three and a half years after operation. This patient's maternal aunt had died from cancer of the breast, and her maternal uncle from cancer of the throat.

5. *Carcinoma of breast of six months' standing ; removal of breast ; well for three years and a quarter, when return took place in scar and opposite breast.*—Bridget W., aged 35, a married childless woman, applied to me on December 21st, 1865, with an open cancerous ulcer occupying the seat of the right breast, which had been removed four years before for carcinomatous tumour of six months' growth. The patient had remained well up to nine months before. When seen, the *left* breast was likewise infiltrated with cancer, and the skin over the chest was covered with tubercles. Four months later the patient was sinking.

6. *Double carcinoma mammæ ; removal of one breast ; return of disease in the opposite ; removal of gland ; death from chest disease* 2½ *years after first operation.*—Miss C., aged 45, consulted me in 1874 with an infiltrating carcinoma of her *right* breast which she had discovered one year. The axillary lymphatic glands were not enlarged. On January 13th, 1875, I excised the breast, but did not explore the axilla. A year later the opposite breast became diseased, and was removed, and within 6 months tubercles rapidly developed over her chest, and she died from chest cancer on July 7th, 1877.

7. *Carcinoma mammæ (double) ; one removed ; second breast involved two years later.*—Mrs. S., aged 46, the mother of one child, which she suckled, came to me with cancer of her *left* breast of six months' standing,

and infiltration of the skin over the breast and of
the axillary lymphatic glands. Two and a half years
previously she had had her *right* breast excised for
cancer of six months' growth. No operation was
advisable. Subsequent history of case unknown ;
but she could not have survived many months.
This patient's parents were alive, and one was 70
years of age. Her paternal grandmother had cancer
of the womb. One paternal uncle had cancer of the
stomach, and a second cancer of the liver.

8. *Cancer of both breasts ; one successfully treated
by injections of acetic acid.*—Mary W., aged 62, the
mother of **ten** children, all of whom she suckled,
although with one twenty years before she had had an
abscess in her right breast. She came to me on April
19th, 1866, with her right breast enlarged from carcinomatous infiltration, the skin over it adherent ; the
nipple retracted, and the axillary glands enlarged.
There had been a lump below the nipple for five years.
I injected acetic acid into the growth **in many** places,
and on many occasions, and the operation was followed
by sloughing of the whole mass, with the skin over it.
The wound subsequently healed, and a good scar
formed. On November 29th she came to me with the
opposite breast the seat of a central tumour. After
this **her** powers soon began to fail.

9. *Carcinoma of both breasts (chronic) ; lived six
years ; no operation.*—Eliza C., aged 72, a single
woman, came under my care on February 6th, 1865,
with a chronic carcinoma of her *right* breast of *five*
years' standing. The whole gland was infiltrated and
hard. The tumour was fixed to the chest, and the
skin over it was puckered and covered with tubercles.
The nipple was retracted and also infiltrated ; the axillary glands were enlarged. With tonics the patient's
health improved, and the enlarged axillary glands
disappeared. June, 1865, the left breast became

involved, the first symptom being some **general thickening** ; by November the infiltration of the gland was very general, but the nipple remained natural. In November the right breast began to ulcerate. January, 1866, cough appeared. April, 1866, the right arm **became** œdematous, the left **breast was** enlarged, and shot-like tubercles appeared in **the skin over the breast.** A few months later the patient died with chest trouble. The disease on the whole had lasted six years.

10. *Cancer of both breasts of three years' standing ; no operation.*—Mary S., aged 58, the mother of one child, aged 24, who, when suckling, so bit the left nipple that it ulcerated off, came under my care on July 26th, 1869, with cancerous infiltration of **both** breasts. Skin over them **adherent and puckered,** and axillary glands likewise swollen. **The** disease began three years before I saw her in **both** breasts **at once.** The patient **sank in the course** of a few **months.**

Treatment of recurrent growths.—In the treatment of these a surgeon's judgment should be guided by the same principles of practice which influenced his decision in the case of the primary operation ; that is, if the recurrent growth stands alone, and can be removed, the operation is justifiable. It may be that the growth, which appeared to be recurrent, was but the continued growth of some of the **elements** of the tumour which had not been removed **at the** original operation ; **and if** so, the prognosis of the case would be more favourable than it would otherwise be. It may be that such recurrence is due either to some local or lymphatic infection, or to both. Under all circumstances, however, if recurrent growths can be removed they should be so treated.

When only such portions of breasts as seemed

affected with scirrhus were removed, what were called
recurrences were frequent.   Such growths were, however, wrongly named, for they would now be called by
surgeons continued growth of pre-existing carcinomatous centres ; and in past years when the lymphatic
glands were neither removed when they were markedly
enlarged, nor looked for by an incision when apparently
free from enlargement, the same remarks are applicable.
True recurrence, when not explicable by either of these
methods, can only be explained by either the appearance
of a fresh focus, or by a dissemination of the local
disease by a more obscure local or lymphatic infection.

Cases unfit for operation.—An operation
on the breast, as in other cases, may be inexpedient
from either local or constitutional causes.   Cases in
which the disease cannot be entirely removed should
be left alone ; as well as those in which the disease
can be wholly removed, but the patient, either from
age or general condition, induced by any cause, is too
feeble to undergo the ordeal of an operation.

Locally, tumours that are fixed firmly to the
pectoral muscle and parts beneath ; tumours complicated with lymphatic glandular troubles which cannot
be eradicated, more especially enlarged glands above
the clavicle ; tumours associated with œdema of the
arm on the affected side, as well as those with
secondary external or internal metastatic growths,
are not to be interfered with.

Cases of brawny carcinoma had better not be
touched ; nor had those of atrophying scirrhous or
other varieties of the disease which are associated
with the presence of tubercles more or less diffused
about the skin.   Cancer *en cuirasse* should never be
interfered with.

To gain a good idea of the effects of operation in
cancer of the breast upon the duration of life, a
statistical comparison of the two classes of cases, that

is of those operated upon and those left alone, must **be made.** In a former page (152) I gave a table of my own, showing the results of operation, and compared it with a second of Sir J. Paget, composed of cases in which no operation was performed.

I give here Gross's statistics upon the **same** question, he being the most recent writer upon the breast.

I have embodied his conclusions in the following table, which tells strongly in favour of operation :

						Not operated upon.	Operated upon.
Died in between **5 and 12** months.						30·8 p.c.	10·5 p.c.
,,	,,	**12**	,,	24	,,	37·7 ,,	33· ,,
,,	,,	24	,,	36	,	11·3 ,,	24· ,,
,,	,,	36	,,	48		9·8 ,,	9·9 ,,
,,	,,	48	,,	60	,,	7·9 ,,	7·9 ,,
,,	,,	60	,,	72	,,	3·4 ,,	5· ,,
,,	,,	After 6 years				1·2 ,,	9·5 ,,
Average duration of life,						27 months.	**39 months.**

(1 year being added to life by the operation).

Velpeau stated in 1864, at the French Academy of Medicine, that out of 250 cases in which patients **had survived** the operation of amputation of the breast, twenty, or one in twelve and a half, had remained free from disease for more than five years, some few having been so for ten or twenty ; and **these** operations must have been what would now **be** called incomplete operations, for before and at the **date named the** lymphatic glands were rarely touched.

Nunn likewise states that one case in thirteen on an average lasts from ten to twenty years.

Compared with these results, **my own** table is very favourable, for at least one-fourth of the cases lived from five to ten years or more. But it must be re- **membered that** in at least a third of these cases **the** major operation of the complete removal of the breast and lymphatic glands was carried out.

I have been unable to make from my own cases, as Gross has made from his, a comparison between cases that have been operated upon, and those that have run their natural course, the difficulties of tracing patients who have not been submitted to operative treatment being greater than in those who have. My own table of operative cases, however, fairly supports Gross in showing the good effects of operation, and I think more forcibly, for in mine at least 14 per cent. lived over six years against his 9·5 per cent., and 25 per cent. lived over five years against his 14·5 per cent.

Life is therefore probably prolonged by operation, and more particularly by what must be described as a complete operation ; indeed, by such a measure, even where it includes the removal of the cancerous breast, with its infected lymphatic glands, a permanent recovery may, in a certain proportion of cases, be looked for. In what exact proportion of cases this result can be secured, it is difficult to say, but if we adopt Gross's method, and assume that when a patient has survived an operation over three years without any local sign or general indication of recurrence, she may be pronounced cured, I do not think his proportion of 9 per cent. of all cases is too high. Recurrence may, however, take place at a date long after three years, consequently Gross's view can only be an average one.

At page 158 a case is recorded in which, after operation, no recurrence took place for twenty-five years, and on looking over my notes I read the case of a woman, aged 47, who was operated upon after the local disease had existed for two years, who remained well for five years when a local recurrence took place, and a second operation was performed ; the second operation being followed, after the lapse of another five years, with a second recurrence which rapidly grew and destroyed life.

Case 2 was that of a woman, aged 50, who, after having been operated upon for a disease of two years' standing, remained well for eleven years ; then a local **return of her trouble appeared,** which was removed, **and two years later she** was **well.**

Case 3, a patient of Mr. Birkett's, who **had her** breast removed in 1851, when 67 years of age ; she remained well **for** twenty-three years, and at the age **of** eighty had cancer of the scar and opposite breast.

Case 4, also a patient of Mr. Birkett's, who had her breast removed at **the age** of 30 ; thirteen years later the disease recurred in **the** scar, when a second operation was performed. Two **years later a** second **recurrence** was followed by a **third operation, and** there the record stops.

Case 5. In 1876 I removed the right breast of **a** woman, aged 48, for carcinoma, with a good result, from whom Mr. Hilton had, ten years before, in 1866, removed the left breast for the same disease.

In neither operation were the axillary glands felt enlarged, nor looked for by an exploratory operation. In 1879 the woman was well.

Gross gives, moreover, a table of forty-seven cases which he regarded as cured.

14 of these had been free from disease after operation for 3 years and some months.

18 had been well from 4 to 6 years and some months.

7	,,	,,	6 to 9 ,, ,,
5	,,	,,	9 to 12 ,, ,,
3	,,	,,	12 to 15 ,, ,,

The average time of **cure was** five years and nine months, and the disease had existed before operation **on an** average of eighteen months.

Mitchell Banks* **more** recently has given us a table of complete operations in which the primary **tumour** and lymphatic glands were freely removed. **Seventeen** out of sixty-four cases are reported **to have been** alive, and to **have** remained free **from** the disease for three years and upwards, five having been well from three to four years ; three from four **to five years** ; **three** from five to six years ; two **for six years** ; two for eight years ; one for twelve years ; and one for thirteen years.

This advantage of operation is further seen when it is undertaken early. **That a woman would have a better chance of gaining the full** advantage of an **operation for cancer when it is** performed before lymphatic infection has taken place, is only what might be expected ; **the** facts of Winiwarter and Oldekop partly prove this to be the case. "Thus, of 136 patients subjected to operation, 43, or 31 per cent., were free from glandular complication, **and their average** duration of life from the first observation **of the dis-**ease to the fatal issue was 52·7 months ; **whereas of** the ninety-three cases in which the lymphatic glands were enlarged and removed with the breast by opera-**tion,** the average duration of life was only 39·3 months, **or 13·4 months** less than the more favoured class. In the uncomplicated cases recurrence of the growth ave-raged eight months, in **the complicated** 1·9 months." The major **or more** complete operative measures appar-ently have done little or nothing in the way of post-poning recurrence of the disease.

Gross adds that nearly one-third of the patients **who had been subjected** to the radical operation **were free from** disease **after the lapse** of **six years,** while of the patients in whom no operation was practised only one survived that period.

Local recurrence after operation. — That

* Harveian Society ; March 3, 1887.

this is to be expected general experience justifies, but
at what period it may be looked for is most uncertain.
Gross reports that of 368 patients operated upon, all
but seventy-two had recurrences ; that is, in four cases
out of five recurrence is to be looked for. It would
have been interesting to know in how many of these
368 cases the complete operation had been carried out.
Winiwarter and Oldekop go further, and by an analysis
of 203 cases demonstrate that :

In 39 cases it occurred within 15 days.
" 50 " " 1 month.
" 38 " between the 1st and 4th month.
" 18 " " 4th " 6th "
" 16 " " 6th " 9th "
" 19 " " 9th " 12th "
" 9 " " 13th " 18th "
" 6 " " 19th " 24th "
" 3 " " 25th " 30th "
" 3 " " 31st " 36th "
" 2 after 3 years.

From this table it is evident that in more than
half the cases a recurrence takes place within three
months of the operation ; in a quarter of the cases recurrence may be looked for between three and twelve
months ; whereas in the remaining fourth the recurrence may be looked for at any time during the three
years. After this time it is not to be expected, and
it is to be hoped that the patient may be regarded as
cured.

The longer the interval between the primary operation and the so-called recurrence, the better the prognosis. The surgeon should always look upon cases of
so-called recurrence within three months as probable
instances in which some portion of the growth, or
some of the lymphatic glands had been left behind ;
in fact, he should regard them as examples of continued growth of some overlooked carcinomatous
centre.

When a cancerous breast is *alone* removed, writes Gross, recurrence is to be expected on an average of 3·1 months; when it, with diseased axillary glands, is taken away, that is, when what is called the complete operation is performed, recurrence of disease averages 7·5 months; the more severe or complete operation apparently not doing much more than postpone recurrence for a few months.

It would seem from Winiwarter, Oldekop, and Henry's statistics, that half the patients that die from cancer do so from metastatic deposits, whilst the other half die from the baneful effects exerted upon the nutrition of the patient without cancerous degeneration of the viscera.

These authors compute the average date of death from metastasis from the first appearance of the disease at 31·2 months, or fifteen months after lymphatic infection; metastasis may occur from five months to eight years.

Metastatic deposits.—To demonstrate the relative frequency of the seats of secondary deposits, Gross has compiled a table from his own cases, added to those of Arnott, Henry Morris, and Clark, published in the 27th volume of the Transactions of the Pathological Society of London, p. 264, in this way tabulating 128 post-mortem examinations, from an analysis of which he shows that secondary tumours were found in the

				Per cent.
Axillary glands	in	115	cases or	89·
Other	,,	30	,,	23·
Lung	,,	28	,,	22 } 45
Pleura	,,	30	,,	23 }
Pericardium	,,	3	,,	2·3
Peritonæum	,,	3	,,	2·3
Brain	,,	3	,,	2·3
Œsophagus	,,	1	,,	0·78
Stomach	,,	5	,,	3·9
Jejunum	.	1	,,	0·78

			Per cent.
Liver glands in	55	cases or	43·
Pancreas	,,	1	,,	0·78
Spleen	,,	3	,,	2·3
Kidney	,,	5	,,	3·9
Adrenal	,,	2	.,	1·5
Ovary	,,	7	,,	5·4
Uterus	,,	2	,,	1 5
Bladder	,,	1	,,	0·78
Bones	,,	9	,,	7·0
Muscles	,,	2	,,	1·5

He then concludes that when left to itself, carcinoma inevitably kills by its baneful consequences as a local disease, or by its remote multiplication.

That about one in six, or 16·7 per cent., of the patients die of the operation itself, but that the risk is not so great as to forbid interference, since it adds twelve months to the life of the patient.

That thorough operations definitely cure 9 per cent. of all patients, or more than half as many more as it destroys.

That the patient is safe from reproduction if three years have elapsed since the operation, and that, finally, recurrence may be delayed for several months, or be prevented altogether, by clearing out the axilla at the same time that the entire breast is removed.

Gross maintains, moreover, that the proper measure is to remove the entire breast and its coverings by a circular incision, search for any outlying lobules that may be disseminated throughout the mammary region, dissect off the fascia of the pectoral muscle, and prolong the outer portion of the incision into the axilla, with a view to its thorough exploration. " Experience shows, first, that the seats of recurrence, or rather farther spread of the disease after operation, are the skin, paramammary fat, remains of the mamma, and glands of the axilla ; and, secondly, that recurrence in the axilla is far more frequent

P—25

after removal of the breast alone than when that cavity was freed of its contents simultaneously with the extirpation of the breast.

That excision of the breast, with the axillary lymphatic glands, is a graver operation than excision of the breast alone, is what might be expected. Thus, an analysis of cases made by Dr. Stettegart* shows that of 264 cases in which the breast and axillary glands were removed, 61 cases, or 23 per cent., died ; while of 130 cases in which the breast alone was removed, 10 cases, or 7·7 per cent., died, the mortality of the complete operations being three times as great as of the incomplete.

This mortality, moreover, is evidently relatively greater, since it must be admitted that the complete method of operating in breast cases has only come into vogue during the last few years, when the treatment of wounds generally has been far more successful than it was in previous years.

The question before us consequently is evidently this : Can the more severe operations on an average show either a definitely longer interval from recurrence, or a larger proportion of substantial cures? If experience can answer either of these questions in the affirmative, the surgeon is then justified in submitting his patient to the graver ordeal; if otherwise, is this the case ?

As regards the recurrence of the disease after operation, Gross has shown that whilst in the minor measure of removal of the breast it is to be looked for on an average in about three months, in what is called the complete method it ensues in about seven and a half months. He, however, at the same time, believes he has shown that life is prolonged by this operation about one year.

In my own practice I have not, as a matter of

* Langenbeck's Archives, Bd. xxiv. ; 1879.

routine, explored every axilla **for** enlarged lymphatic glands, and particularly in thin women, in whom, under an anæsthetic, the condition of the axillary glands **can** fairly be ascertained by manipulation. I have, however, **always done so** when enlarged glands could be made out, **and when** manipulation was difficult and its conclusions uncertain. I have had no reason to find **fault** with the results of **my** practice.

Out of 56 complete operations, private and public, performed, with all the advantages of improved wound treatment, I have lost 4, or 1 in every 14 cases, or 7 per cent. ; whereas, on referring to former times, when the condition **of the** axillary glands was only surgically considered when their enlargement was manifest, and wound treatment was as a whole **less** successful, I lost **8 out of** 120 cases, **or 1 in 15** cases, or 6·6 per **cent.**

It is true that the causes **of** death in both classes **of** cases **can in a** measure **be** explained away, **and** evidence brought forward to suggest that the operation in at least half the cases had little or nothing to do with the result. I have no intention, however, to do this, **as the ultimate** issue of the question must turn upon the main facts. Could a carcinomatous **tumour** always be removed in its early infiltrating stage, that is, when the disease **is local,** cures may be hoped for. When it is taken away after it has advanced and spread by lymphatic infection, the chances **of a** complete cure are remote. Other things being equal, **free excision** is more likely **to** cure. But are they equal **if the** mortality of the complete measure is in excess ?

When this operation of **removal is called** for, it should, therefore, as a rule, be complete ; that is, not **only** the tumour should be removed, but the whole mammary gland in which it is placed ; with all affected **or** possibly affected skin and fat about it ; but **the** practice of exploring an axilla as a measure of routine should not be followed, since it is without doubt a graver measure to the patient, **and the advantage** which it in theory possesses is not yet proved.

The removal of a breast for cancer **is often expe-**dient and necessary for purposes **of** relief rather than of cure; and **a** breast may often be removed with this object **when** the exploration of the axilla is neither called for nor justifiable. The surgeon's judgment in this matter is wanted in each case as it may come before him, and he is not to be governed by a dogmatic rule which experience soon proves is not a safe **one.**

A series of cases could readily be given by one surgeon **to show** the advantages of the complete measure; whilst another, or possibly the same **sur-**geon, could give another series equally striking, to demonstrate the value of the so-called incomplete. **A** measure which is right and justifiable in one **case may** be wrong and unjustifiable in another. An operation which offers a prospect of cure, and removes what is a source of mental and bodily worry, should always be performed in preference to any partial measure, unless the general condition of the patient is such as to lead **the surgeon** to believe that any extra risk is incapable of being borne with safety. The advantages of **the** graver operation are not, however, so certain as to justify increased hazard to life.

On excision of the breast.—There is, in average cases, no **great** danger attending excision of the breast beyond **that** which accompanies any, even the smallest, operation. **In** feeble and aged subjects **there** is, however, much risk, and a certain proportion **of** cases may be expected to sink after the operation **from** pyæmia, erysipelas, or visceral disease, since these contingencies attend any operation or wound. From **my notes of** hospital operations on the breast for cancer, or rather of 200 consecutive cases, only 10 died from pyæmia, pneumonia, or erysipelas, the mortality being but five per cent., and these causes are yearly diminishing. Out of 176 cases in my own practice the mortality was 12, or 7 per cent. Gross gives a mortality **of 16 per cent.**

In operating for cancer it is unquestionably the wisest course to excise the whole gland. It is well not to be over-anxious about preserving too much integument, and if any doubt exists as to its perfect healthiness the suspected portion had better be excised. When enlarged lymphatic glands clearly exist they should be taken away, and in most other cases where their existence is suspected the axilla should be explored. It is always important, when dissecting out the tumour, to keep clear of all diseased tissues, and, in fat subjects, to take away much of the fat, since there is good reason to believe that an early return of the affection is too often to be explained by want of attention to these points. When the pectoral muscle is infiltrated it must be freely excised. The pectoral fascia should be closely inspected, since in it small cancerous nodules are often found, which, if left, would have been the centre of new growths.

The operation.—In the removal of a breast the patient should be placed on her back with the shoulder of the affected side raised by a pillow, and the arm drawn out at a right angle to the body.

The incision should be elliptical, and made in a line parallel with the fibres of the pectoral muscle. When the skin is diseased and has to be removed, the incision may have to be circular, and bleeding is to be controlled by the pressure of the fingers of an assistant, or by torsion. The outer or pectoral incision should first be made, and carried well down to the free border of the pectoral muscle, the definite form of which is the best and surest guide to the

base of the gland. The inner or sternal incision may
then follow. The whole tumour should then be excised and dissected away, a few touches of the scalpel
clearing it off the pectoral muscle. The axillary angle
of the tumour should be divided last, as it usually

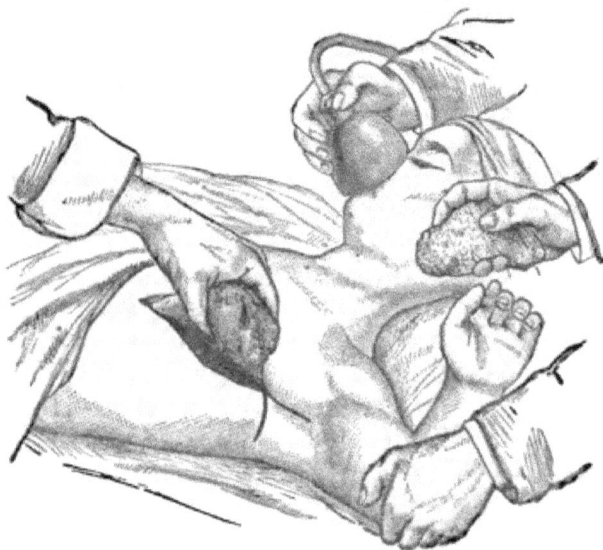

Fig. 6.—Excision of the Breast.

contains the lymphatic cords and chief vessels that
supply the gland. The incision can be extended upwards into the axilla along these cords to explore
or remove the glands. When this is done an
incision at right angles to the wound backwards
is good for drainage purposes (Fig. 6). All
bleeding vessels should be twisted, the surface of the
wound cleaned with iodine water (ʒii. ad Oi.), carbolic
acid lotion (ʒiv. ad Oi.), bichloride of mercury (gr. x.
ad Oi.), or chloride of zinc (gr. xx. ad ʒi.), its edges
well adjusted, a drainage tube introduced at the most

dependent part, and steady pressure applied by means of pads of antiseptic lint, gauze, or cotton wool, the wound being treated on ordinary principles.

The axillary glands should be removed more by enucleation with the fingers and a **blunt dissector** than by the knife; by care and a little patience this may usually be accomplished with safety. The surgeon must be careful not to draw too much upon the axillary mass where such exists, for by so doing he will probably drag the large vessels, and more particularly the veins, out of their position, and expose them to injury. All parts that are cut through should be ligatured first, and then divided in the distal side of the ligature. Banks advises that the axillary **vein** should be first exposed, after which no danger of wounding it will be experienced. Should the vein be wounded it should be tied.

Beyond satisfying **myself that the drainage of the** wound is satisfactory, I rarely **remove the first dressing** for three days, when, in favourable cases, what part of the wound was wanted to heal by quick union, has repaired, and the open surface is granulating. **This** I wash with iodine water, and dress with lint or iodoform gauze, **soaked in a** mixture of one part terebene and three parts olive oil, covering in the whole with a sheet of Gamgee tissue. By these simple measures repair is carried out.

The essential point the surgeon has to observe in **the** operation is *thoroughness;* with this, success may be hoped for **and** reasonably expected ; without it failure and disappointment must be the result.

Inadequate **and too late** operations have doubtless been the cause of many of our failures, and these again explain why it is that some practitioners refrain from advising operative interference, and consequently prevent their patients from seeking early surgical help.

When the profession, as a body, is led to look upon cancer of the breast in its origin as a local disease, and consequently one to be treated early; when surgeons are bold enough to advise, in early doubtful cases, an exploratory incision into the growth with a view to its excision should the disease be found to be carcinomatous, better results will be obtained. Excision is likely to be most successful when applied in the stage which precedes that of lymphatic infection, and when the cancer may be regarded as a local affection. Those who come after us will doubtless have to record better and more encouraging results than I can adduce in these pages.

On the treatment by caustics.—This form of treatment has its place in surgery, but it is not to be considered in competition with excision, since by excision alone can the whole disease be taken away with the greatest certainty, and in the shortest possible time.

Caustics are, however, useful in open cancerous ulcers, that cannot be excised, with the view of checking growth, preventing fœtor, lessening discharge, and diminishing hæmorrhage.

A paste consisting of one ounce of a liquid extract of the sanguinaria canadensis, made by boiling down a decoction of the root, in which the same quantity of· the chloride of zinc is dissolved, and then mixed with two ounces of the extract of stramonium, is very good ; and so is the paste used at the Middlesex Hospital, made by mixing chloride of zinc and boiled starch with laudanum, till it reaches the consistence of honey.

At the Middlesex Hospital, where this method of treatment is practised more frequently than elsewhere, Mr. H. Morris tells me it is carried out as follows : The whole cutaneous surface of the breast to be removed is first destroyed by the action of fuming

nitric acid applied by a glass brush ; and in doing
this, care is called for to guard against the effects of
any running of the acid over the surrounding parts
by the application of some grease, and having at hand
an alkaline lotion to neutralise the acid.

When the skin has been destroyed, as indicated
by its parchment-like feel and bloodless aspect,
multiple incisions well into its substance should
be made radiating from the nipple ; the surgeon taking
care not to cut so deeply as to draw blood.

Into the furrows thus made in the dead skin
strips of finely prepared lint should then be pressed,
the caustic paste having been previously spread upon
and well pressed into its meshes, and all redundant
paste scraped off ; the object of this being to prevent
the paste running over healthy tissue and causing
distress. The breast should then be covered with cotton
wool and a bandage.

In a day or two, when the caustic has done its
work, the strips of caustic lint should be removed, the
furrows in the breast deepened by fresh incisions, and
strips of lint, prepared as before, applied. The same
care being observed by the surgeon in his incisions
not to draw blood.

By a repetition of this process every two or three
days a breast may be removed in about three or four
weeks, and, it is said, with but little pain.

Upon this point I am unable to speak from
experience, as I have never removed a whole breast
by this means. In some cases, painting an ulcerated
surface over with liquefied carbolic acid is of great
value, the acid acting as a caustic, disinfector, and
anæsthetic at the same time.

Of late years Esmarch's powder has been much
advocated, and I think with justice. It is said to be
painless. It is made by mixing one drachm of arse-
nious acid with the same quantity of sulphate of

morphia, eight drachms of calomel, and six ounces of powdered gum arabic. The surface of the sore is to be thickly sprinkled daily with this powder till a crust forms, and repeated if necessary.

As an ordinary application to a cancerous sore where these severe caustics are not employed, I find an ointment made of two drachms of resorcin with six of vaseline is very good. I have thought that it dissolved the epithelial structures, and consequently tended towards a cure. I have, in a few cases, used resorcin as a powder, dusted over a cancerous surface, and with advantage. I have thought it tended to bring about its disintegration.

Injections of acetic acid into cancerous growths cannot be said to be quite useless, since, in one case already quoted (page 216) the injections brought the mass away ; but their effects are so uncertain and unsatisfactory, that they are not to be relied upon, and by their use valuable time is lost.

When no local treatment of any special character is employed, and the disease must be left to take its course, pain must be assuaged by the internal or hypodermic use of morphia in one of its forms, the bimeconate being the best ; or the local application of a drachm of the extract of belladonna, rubbed down with one ounce of glycerine, or of the extract of stramonium ; or of an ointment containing a tenth part of cocaine. As a painless deodoriser, Gross speaks highly of a mixture of five grains of chloral hydrate to one ounce of vaseline.

Should great activity be displayed in the local disease as indicated by extreme capillary injection and warmth, the local use of cold gives comfort and checks growth. This may be effected by means of one of Leiter's metallic coils or ice bags, and when these are inapplicable, by the employment of the usual lead lotion mixed with opium.

In foul cancerous ulcers a powder of fine **iodoform**
or iodol mixed with boracic acid, in the proportion of
one to four, is to be recommended ; and in very painful
sores cocaine **used as an** ointment **or** in solution, **of
about** half **a drachm or** drachm to an **ounce, gives
comfort.   Hamamelis.has** been strongly **advocated as**
an application **to open** cancer.   **I have,** however, **failed**
to find its advantages.

Should **local** bleeding occur, a pledget of **lint,**
soaked in a concentrated solution of alum, or of dried
cotton wool which has been steeped in a solution of
the perchloride of iron, will probably suffice to check
it.   The glycerine of tannic acid, applied on lint, **is**
also under these circumstances useful.

CHAPTER XV.

A SERIES OF CASES   ILLUSTRATING   THE   CLINICAL
SYMPTOMS AND TREATMENT OF CARCINOMA OF THE
BREAST.   EACH CASE ILLUSTRATES A POINT.

1. *Carcinoma of the breast, unrecognised in its early
infiltrating stage ; subsequent excision and recovery.*
—**Mary A.** C., a healthy looking single servant, aged
37, came **under** my care on March 31st, 1877, with
some swelling of **her** left breast. **Her** mother, maternal
grandmother, and aunt all died **of cancer of the**
womb ; father alive and **well.**   For three months **she**
had had shooting pains in her left breast, and **for one**
month had noticed that it had become hard and
tender.   She thought also the nipple was retracting.
When I saw her, one lobe of the gland was certainly

slightly fuller and harder than the other; it was likewise very tender to the touch. The nipple was not
to any degree more retracted than that of the opposite
breast. The skin over the breast was natural, and
the axillary glands free. I was disposed to look upon
the case as one of irritable mamma, and treated it
accordingly. In a year she reappeared with a typical
carcinomatous tumour, as indicated by a hard nodulated swelling in the position of the former tender
and slightly hardened lobe; adherent puckered skin
over the tumour, well-marked retracted nipple, and
enlarged axillary glands.

I then removed the breast, tumour, and axillary
glands, and a good convalescence followed. Four
years later this patient was well.

Remarks.—In this case an exploratory incision into
the growth when first seen would have saved delay
and a surgical error, although the subsequent progress
of the case has been one of hope. It tends likewise
to support the practice of making an exploratory
incision into any doubtful tumour of the breast when
first coming under observation.

2. *Carcinomatous tumour attached to margin of
breast, simulating adenoma; excision of tumour; patient
well three years subsequently.*—Catherine C., a married
woman, 33 years of age, the mother of five children,
came under my care in June, 1874, with a tumour in
her left breast, which had appeared six months previously, when she had given up suckling her last
child. The tumour had grown slowly since, and when
seen was the size of a hen's egg. It was situated at
the sternal margin of the left breast, and was hard,
nodulated, and very movable. The skin over the
tumour was free and natural; the nipple was normal;
axillary glands natural. The growth was regarded
as an adeno-fibroma and removed, the breast itself
being left. A good recovery followed the operation,

and the patient was well three years subsequently. A section of the tumour, and a careful examination of its structure, **proved that it** was an example of scirrhous carcinoma.

I regretted at **the** time that the whole breast had not been **taken away.** The last report of her **case** was, **however, that she** was well three years later, and **under such circumstances** the feeling of regret has disappeared. **The tumour must** have been, when removed, **a** local disease.

3. *Carcinoma of breast ; dimpling of skin ; excision of growth ; recovery.*—Mary A. T., a healthy looking, childless, married woman, aged 30, came under my care on April 11th, 1881, with some thickening of the clavicular lobe of her left breast, which had **been** coming on for four months, after a blow. **The tumour was** ill defined and firm, but **not so hard as scirrhus** generally is ; indeed, **from its** hardness alone **no** diagnosis **could** be made, particularly **as** the nipple was natural and there were no enlarged axillary glands. The skin over the tumour was, however, slightly dimpled ; with this symptom the diagnosis **was** made, and **the** breast and tumour first incised **and** then excised. **The** disease was clearly on section carcinomatous. A good recovery followed.

In this case the value of the dimpling of **the** skin **over** a doubtful tumour was well displayed, **since it** enabled a definite diagnosis to be made as **to the nature** of the growth, and active treatment **to be based upon it.** Had this dimpling not existed, **an exploratory** incision into the tumour, for diagnostic purposes, would **have** been called for, prior to its excision.

4. *Carcinoma of breast ; painless progress ; tumour and dimpling of skin ; nipple natural ; operation and recovery.*—Susan V., a childless married woman, aged **58, came** under my care in April, 1873, with a hard,

nodulated tumour of the upper segment of the right breast, about 2½ inches in diameter, which had been slowly growing for two and a half years, and absolutely without pain. The nipple of the gland was natural; the axillary glands were not enlarged, but the skin over the tumour was dimpled. The breast and tumour were excised, and a good recovery ensued.

5. *Carcinoma of breast and axillary glands; skin over tumour and nipple natural; excision of disease; recovery; five years later the patient was well.*—Eliza H., aged 42, the mother of four children, all of whom she suckled without trouble, came under my care on March 14th, 1872, with a tumour the size and shape of a small orange, in the axillary border of her left breast, and a mass of enlarged axillary glands beneath the anterior border of the pectoral muscle. The tumour and breast moved freely over the parts beneath; the skin over the tumour was normal and quite free; the nipple was natural. On March 22nd the breast was removed and the axilla was cleared out. Recovery followed, the woman being well five years subsequently.

6. *Carcinoma of the breast; attention drawn to part by retracting nipple, not from pain; operation.*—Eliza J., a thin, healthy looking woman, aged 60, the mother of six children, came under my care on January 6th, 1883, for a tumour in her left breast. She had had no pain, or even discomfort, in her breast, and her attention was first drawn to it by a nurse who, when dressing her, observed that the nipple was retracting. This led her to see her medical man, who sent her to me, when I found a small typical carcinoma fibrosum, as indicated by a stony nodular infiltration of the whole breast, with retracted nipple, and puckered integument over the tumour. The lymphatic glands were not involved. As the patient was

thin, and apparently a good subject for an operation, and the disease seemed to be quite local, I excised the gland, and a good recovery followed. The **axilla was** not explored. The patient was seen eighteen **months** later, and **she was well.**

7. *Carcinoma attacking breast at the seat of an old abscess; nipple retracted; skin and lymphatic glands uninvolved; operation; recovery.*—Emma C., aged 59, the mother of two children (the youngest twenty-eight **years** of age) came under my care on May 17th, 1880, **with a** tumour in her left breast. When suckling **her last child,** twenty-eight years previously, she had **an abscess in her** left breast. Eight months ago she **found a lump situated** in the breast beneath the scar **of** the old abscess, and this lump **has** steadily increased, although without pain. At present there is a hard, nodulated swelling, **the size of an** orange, upon **the** upper and sternal **segment of the left** breast. **The tumour** is clearly in the gland, **which moves with it** over **the deep** parts. The skin **over the tumour is** healthy **and** movable, except where the old scar of the abscess exists. The nipple **is** much retracted; **the lymphatic glands are apparently** uninvolved. May 21, **the breast and tumour were** removed, and a good recovery followed. **The tumour** was a good specimen of carcinoma. **I have been unable** to trace the later history of this case.

8. *A carcinomatous tumour attached to the upper border of right breast following a blow; excision of tumour; patient well four years later.*—Eleanor C., a childless widow, aged 40, came under my **care in March, 1874,** with a tumour **four** inches square, **attached apparently** to the upper margin of **the** right breast, **although** not in the breast, which **seemed** quite normal **in all** ways. **The** tumour had been growing **for thirteen** months, **and** had followed a blow received **one** month previously. **It** had grown slowly, and when seen **was**

somewhat fixed, although movable upon the parts
beneath ; the skin over it was infiltrated, red, smooth,
and part of the tumour. The axillary glands were
free. On March 17th I excised the tumour and skin
over it, leaving the breast, which did not seem to be
involved. A good recovery followed, and four years
later this patient was still well.

9. *Carcinoma of left breast; sloughing ; operation
and removal of axillary glands ; relief.*—Mary B. J.,
aged 55, the mother of two children, came under my
care on October 3rd, 1883, with a large fungating,
bleeding, carcinomatous growth, involving the nipple
and the whole of the left breast ; also the axillary
glands. It was movable upon the deeper structures.
The disease had existed for six years, had grown
slowly for five, more rapidly for one. Nine months
before admission into Guy's Hospital it began to
break down, and for three months it had bled. On
October 9th I removed the whole mass, with the
axillary glands, and a good convalescence followed.
Some months later she was well, and had greatly
improved in her general condition.

10. *Carcinoma of both breasts; excision of both at
an interval of two and a half years ; convalescence.*—
Harriet W., aged 60, a governess, came under my care
in March, 1884, with a well-marked carcinoma of her
left breast, of eight months' standing. It was the size
of a walnut, and was placed on the axillary lobe of the
gland. It was movable over the deep parts, and the
skin over it was neither fixed nor dimpled. The nipple
was retracted ; axillary glands free. Two and a half
years previously she had lost her right breast for an
open cancer of many months' standing. On March
28th I removed the tumour and breast, and a good
recovery followed. The breast had almost disappeared
by atrophy.

11. *Carcinoma of the breast, axillary glands, and*

skin, with discharge from nipple ; operation ; recovery.
—Mary M., a married woman, aged 48, the mother
of three children, all of whom she had suckled, came
under my care in September, 1874. Her grandmother had died of tumour of her breast at the age
of 79. Her mother was alive and well. The patient
had discovered a lump in her right breast about one
year before admission, and at the same time a swelling
in her right arm-pit. There was no pain in either
swelling for some months, but for about six months
pain appeared as the swellings increased in size. When
seen there was a dense, hard, nodular swelling in the
upper half of her right breast. The skin over the
swelling was pitted like the rind of an orange, and
adherent to the tumour beneath. The nipple was
drawn inwards, and on pressure over the tumour a
clear brown fluid exuded from the nipple, which,
when tested, was highly albuminous. In the skin
above the tumour was a single, red, hard tubercle,
which the patient said had followed a poultice ; the
axillary glands were enlarged, but movable.

On October 6 the tumour with the breast, skin
over it, and axillary glands, were removed, a good recovery following. Sixteen months later the woman
was well. The tumour was a good example of scirrhous
cancer.

12. *Rapidly growing tuberous carcinoma of the
breast with all the well-marked symptoms ; operation ;
recovery.*—Emma P., a healthy looking woman, aged 39,
the mother of four children, all of whom she suckled
without trouble, came under my care on March 12,
1879, with a hard, nodulated tumour, measuring five
inches by three, in the outer side of her right breast,
to which the skin over it was attached by infiltration.
The nipple was likewise retracted, and the axillary
lymphatic glands enlarged. The tumour was first
discovered only three months before, when it was the

Q—25

size of a walnut, the breast and nipple at that time appearing to be otherwise natural.

On March 18 the **breast** tumour and axillary glands were cleared away, the patient making a good recovery. The specimen removed **was** an excellent example of rapidly growing soft cancer. It was soft, highly vascular, and lobulated, although clearly infiltrating the breast structure. No subsequent history **of this** case can **be** obtained.

13. *Carcinoma of the breast ; slow, and later on, rapid growth ; radical treatment with good result.*— Mary M., **aged 57,** the mother of eight children, the youngest **being 20** years of age, came under my care in April, 1876, **with a** hard, nodular tumour, measuring **four** by five **inches** in diameter on the outer and lower segments of her right breast. The skin over the tumour was "puckered," the nipple retracted, and the axillary lymphatic glands enlarged. The tumour had been growing for nearly three years, although for **two and** a half years its increase had been slow ; since **then it** has grown rapidly. At first it **was the** seat **of** occasional pain ; for the last six months the pain has been acute. The nipple was unaffected up to three months ago, when it began to retract ; at the same time the skin over the tumour became dimpled, and **later** on puckered to **the** parts beneath. The axillary glands became likewise affected.

On May 2 the tumour, **with** the breast and axillary glands, **were** cleared **away, and** a good result followed. The patient was well **a** year later. The history of this case illustrates two different conditions. The first **one** of slow and **inactive** growth ; the second of rapid **and** active increase of the original tumour, and local **and** lymphatic infection of other parts.

14. *Carcinoma of the breast accompanied with severe pain ; enlargement and resolution of lymphatic glands ; operation ; recovery ; well three years later.*—Ann N.,

aged 51, the mother of nine children, came under my care on February 10, 1882, with a tumour the size of an orange, in the upper half of her left breast. It had been growing for three years, and for the **first** two and a half years gave but little trouble ; for **six months it had** increased rapidly, and **was** the **seat of severe pain.** Three months previously the axillary glands enlarged, but they subsided under treatment. On admission the tumour is very hard ; it has a **smooth** outline, and the skin over it is " puckered." The nipple is retracted. No enlarged lymphatic glands are to be felt. On February 14 the breast and tumour **were** removed, the axilla was explored without result, and the case did well.

The tumour on section presented in its centre all the features of true scirrhus ; around its periphery it formed a good example of the **more** rapidly growing and softer medullary cancer ; **the tumour** exhibited at the same time the microscopical **ap**pearances of the slow and rapidly growing varieties of carcinomatous disease. In 1886 this patient was **well.**

15. *Tuberous carcinoma of breast with infiltrated* **skin***; natural nipple ; enlarged axillary glands, following a blow, associated with albuminuria ; no operation.* —Maria R., a healthy looking old woman, aged 72, came under my care in April, 1884, with a tumour in **her right** breast. She was a married woman, and had had eight children, all of whom she had suckled without trouble. She had always enjoyed good health. Five years ago she struck her breast **severely** against the corner of a washing tub ; swelling and pain followed, and when the immediate effects of the injury had subsided, a lump remained. Six months ago this lump suddenly began to enlarge and become the seat **of** pain ; for this she sought advice. Her family history was good.

When seen a tumour the size of a tea-cup occupied the upper and outer segments of the right breast. The skin over it was part of the tumour, and was red from congestion, fixed to the growth, and hard. It was in parts slightly œdematous. The nipple was natural. The axillary glands slightly enlarged. The breast and tumour moved freely over the pectoral muscle. Pulse good. Urine, specific gravity 1012, contained albumen. Under these circumstances no operation was advised.

16. *Carcinoma of the breast with tuberculated skin and carcinoma of spine ; no operation justifiable.*— Eliza R., aged 44, a single woman, came under my care on May 1, 1877, with a tumour about two inches by one and a half in the right breast, which had been growing for about five months. When first discovered it was the seat of an occasional shooting pain. The tumour was hard and nodulated, and clearly formed part of the breast. It moved freely with the breast over the deeper parts. The skin over it was "puckered," and in it were many pea-like tubercular cancerous infiltrations. The nipple was retracting, and the axillary glands were unaffected. About six weeks before admission she became lame in the right leg, and pain passed from the loins down the leg. This so increased in severity that she was unable to walk, and sometimes in her sleep her right leg started.

On admission she was unable to sit up without experiencing great pain in her back from the mid-dorsal region to the sacrum, and there was tenderness on pressure over this region, and over the right sciatic nerve. She could flex the right thigh partially, but she could not lift the leg from the bed. There was no loss of sensation in the limb. With these symptoms it was assumed that secondary disease existed in the spine, and no operation was advised.

17. *Carcinoma of the breast; tubercular infiltration*

of the skin; ulceration of **the skin** *; no operation justifiable.*—Eliza G., a fairly healthy looking woman, aged 53, the mother of two children, came under my care on June 20, 1877, with a **hard scirrhous tumour** two and a half inches in diameter, **in the axillary lobe** of her left breast, which had been growing **for one year.** The skin over the tumour, breast, and sternum, **was** mottled red, as if inflamed, and to the touch **was** tuberculated, and irregularly infiltrated with carcinomatous elements. At one spot the skin was ulcerated from the breaking down of the skin cancer. The breast was fixed to the deeper parts, clearly from the extension inwards of the disease. The axillary **glands were** likewise infiltrated. **The** nipple was not **retracted.**

18. *Carcinoma of the breast of a woman, aged* **25,** *in two separate masses; operation refused.*—Mrs. **S.,** aged **25,** discovered, in January, 1872, a **small lump in** the upper half of her right breast, but as it gave no **pain** she disregarded it. She had been married **seven years,** but had not been pregnant. When seen in February, **1873,** one year after the discovery of the tumour, the upper half of the right breast was occupied with **a** nodulated mass the size of an orange, which evidently infiltrated the gland with the skin over **it,** which was puckered. Below the nipple there was a second tumour the size of a walnut, which was unconnected with the first. The skin over it was dimpled. **The** nipple **was** retracting, and the axillary glands enlarged. Operation was refused.

19. *Infiltration of breast following the disappearance of cystic formation ; cupping of skin from scar of abscess which existed twenty-two years before.*—Susan K., aged 48, the mother of one child twenty-two years previously, when she had an abscess in her right breast, came to **me on** November 27, 1865, with great thickening and enlargement of her right breast, associated with pain.

The integument over it on the outer side was drawn down to the gland (dimpled). Two years previously this woman had been under my care for two swellings, which were taken to have been cysts, in the same breast, which had disappeared. November 5, 1866, the left breast became indurated like the right, and this rapidly developed whilst the right remained stationary ; nodules appeared also upon the skin. Operation refused.

The cupping of the skin was due to the contraction of the scar of the abscess which had existed twenty-two years before, the enlargement of the breast making the depression more marked.

CHAPTER XVI.

CYSTIC TUMOURS OF THE BREAST.

WE now come to what is at once the most interesting, and at the same time the most perplexing section of the whole range of diseases of the breast, since it is cystic disease of the gland that, to my mind, makes any pathological arrangement in such a work as this of little utility. It is on this account that I determined in these pages to consider breast diseases purely by the light of clinical experience.

As I have already urged, there are those who, at the present day, treat of cystic disease of the breast as one of the nature of sarcoma. I, on the other hand, contend that this grouping is not right, under the belief that cysts are a part of the pattern or architecture of most tumours of the breast, whatever be their nature, save and excepting scirrhus, and even with this disease cysts are found.

Thus we clinically meet with simple cysts, cystic

adeno-fibromata, myxomata, sarcomata, carcinomata, and what not, and it is left, as yet, entirely to clinical experience to say what is the natural tendency of cystic disease.

I shall, in the following pages, describe cases of all these varieties of cystic disease, and I shall, I trust, show that the greater proportion by far of all these cases is, so far as I can learn, cured by free removal; that is to say, that recurrence after such treatment is exceptional.

The bearing and importance of this fact are obvious, for if it be so, there is much reason for insisting upon the restriction of the term sarcoma at the bedside.

Under this heading I propose to group all tumours of the mammary glands made up of cysts, whether those cysts are multiple or single, whether they contain fluid alone, or are more or less filled with adeno-fibromatous, adeno-sarcomatous, or adeno-carcinomatous intracystic growths.

I shall not stop to inquire into the different methods by which the cysts are formed, since I believe it to be impossible *clinically* to make out, otherwise than in exceptional cases, whether the cysts are due to an obstruction of a galactophorous main or branch duct (retention cyst); or to the exudation of fluid into the intercellular spaces of the connective tissue of the breast (lacunar cysts); or to the growth of papules into a duct or acinus from the connective tissue of the breast and their subsequent junction. Neither can the surgeon speak with certainty in all cases as to the absence or presence of an intracystic growth, nor as to the nature of the growth.

In a large proportion of cases in which the cyst contains solid growth, a diagnosis of its nature can, however, be made, but the diagnosis is then more determined by signs and symptoms other than those associated with the presence of a cyst. To these points attention will be drawn later on.

I would, however, for the sake of convenience and possibly of clearness, divide the subject up into certain sections, and consider:

1st. The cystic degenerations of the breast as met with in the old, as well as in glands which have long ceased to be active; involution cysts as they are called (Plate VIII. Fig. 5).

2nd. Cystic tumours of the gland, single or multiple, of glandular, duct, or connective tissue formation *without* intracystic growths (Plate VII. Fig. 1).

And, 3rdly, cystic tumours of the breast of whatever kind in which papillomatous, adenomatous, sarcomatous, or carcinomatous intracystic growths are present (Plate VII. Figs. 2, 3, and 4).

These three classes will be considered in order.

GROUP I.—*The cystic degeneration of the gland as met with either in the old or obsolete gland.*

This condition must, without doubt, be recognised, for it is found in the breasts either of the old, or of those in whom the glands have ceased to be active, and are, as it were, undergoing involution changes. Meaning by this term involution change, an irregular form of atrophy, a condition to be expected in any tubular or acinous gland, and, as a fact, one well known in the kidney, the thyroid, and occasionally in the pancreas.

It may be that a single lobe or lobule is undergoing this special form of degeneration, but more probably the change will have affected the whole gland, which, on dissection, will be found to be made up of innumerable small cysts, varying from the size of a hemp-seed (the more common size) to that of a pea, whilst in exceptional cases these dimensions, in isolated cysts, may be considerably exceeded.

PLATE VII.

Fig. 1.

Fig. 2.

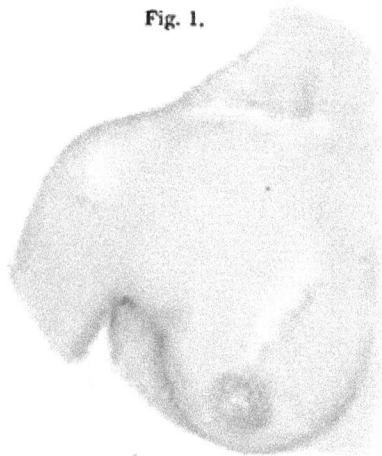

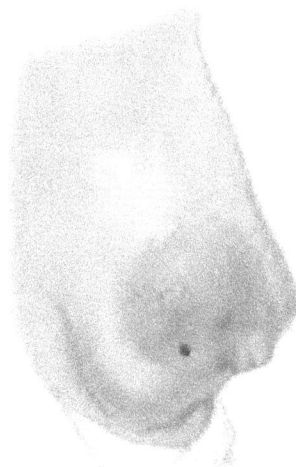

Fig. 3.

Fig. 4.

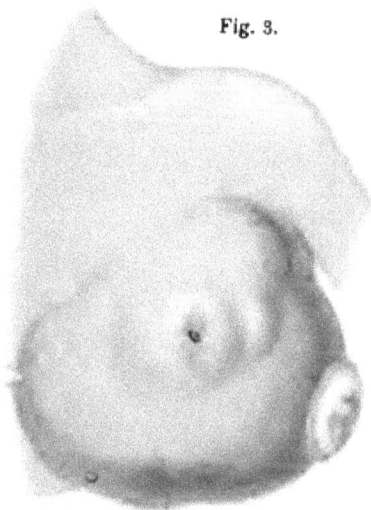

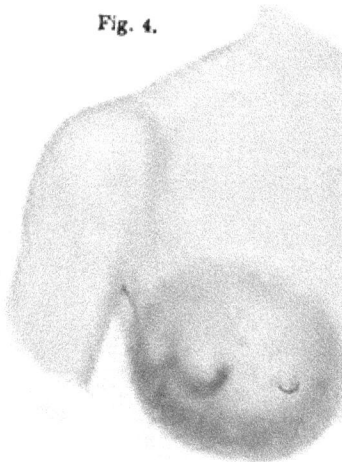

CYSTIC DISEASE.

1. Cystic Disease with discharging Nipple.   2. Cystic Sarcoma.
3. Cystic Sarcoma bursting through Coverings.   4. Cystic Carcinoma.

The cysts appear more as a series of dilatations of the lactiferous ducts (varicose ducts) than of the gland structure, although in certain cases both ducts and gland structure are involved.

On dissecting these cases and on separating the ducts, or rather unravelling them, the gland as a whole may be involved, and in a dissection I made many years ago, the breast appeared, when suspended by the nipple, to be made up of strings of small cysts connected together by the main and branch ducts. Some of the cysts had direct tubular communication with the neighbouring cysts, as proved by the passage of bristles through the ducts from one cyst to another, or by the fact that many of the cysts and ducts could be inflated by means of a fine tube introduced into one or other of the nipple ducts. In many cases, however, this direct communication cannot be made out.

These cysts, when unopened, have usually a greenish or black appearance, and when opened they are found to contain a blackish viscid or mucoid fluid, more or less coagulable by heat, and mixed at times with fat and epithelial elements, such being the products of degenerating epithelium.

The breast, thus affected, feels on manipulation in thin subjects as a coarse gland, with here and there a pea-like tumour, whilst in fat women the change often cannot be clinically recognised.

This change is rarely associated with pain, or with a greater amount than can be described as uneasiness. There is seldom any discharge from the nipple in these cases, although there may be at times. Why there should not be a discharge always can only be explained by the fact that the ducts become obstructed, and probably by their epithelial elements.

In exceptional cases some enlargement takes place of a single cyst or group of cysts, and under such

circumstances a cystic tumour is formed, for which
surgical advice is sought. I have seen many of
these cases. In some the cyst was single, and relief
was given by simply drawing off the fluid by means
of a small trocar and cannula ; in other cases two or
more cysts existed ; in a few they were still more
numerous and showed signs of activity, so that excision
of the affected lobule or whole gland was required.

In not a few cases the cysts have become the seat
of intracystic, sarcomatous, adenomatous, or cancerous growths, for which extirpation of the whole gland
has been necessary. To these attention will be drawn
later on.

What I would wish now to be recognised is the
fact that there is in the mammary glands which have
long ceased to be active, and in those of women past
child-bearing, a cystic degeneration of their ducts and
glandular structure which may simply remain as such
or take on active changes. Under the latter circumstance, a cyst or many cysts may either enlarge so as
mechanically to cause trouble or become the seat of
intracystic, adenomatous, sarcomatous, or carcinomatous growth which will run the usual course of such
tumours.

In the two following cases this cystic involution
change is well illustrated. In the first the whole breast
was excised, and a good result ensued. In the second
some of the cysts were relieved by tapping, when the
remainder disappeared. This favourable result is one
that not infrequently follows an expectant treatment,
but it is not in the majority of cases to be expected.
The hope of obtaining such a success by leaving things
alone, should not induce the surgeon to withhold his
hand in any case in which the breast is clearly enlarging,
although it may be sufficient to sanction a watching
line of treatment where the progress of the trouble is
slow.

Case 1. *Involution cysts of breast which had long ceased to be active, simulating cancer; excision of gland; recovery.*—Mrs. T., aged 52, the mother of many children, all of which she suckled, consulted me in May, 1885, for some affection of her right breast, which had been slowly **coming on** for about six months. **It began as an enlargement of the axillary lobe, and later on as** a swelling **of the whole breast. At times there was** a discharge of a clear **fluid from** the nipple **and** some slight pain.

When seen the whole breast was coarsely enlarged and indurated; one lobe, **the** axillary, seemed to be generally infiltrated with some new material; the other lobes were full **of** nodules which appeared to **be** cystic. The nipple **was** natural, but pressure **upon** the breast caused from **it a** slight discharge of **a serous** fluid. The skin **over the breast was healthy, as were the** lymphatic glands.

I regarded the case **as one** of carcinoma, with cystic degeneration of the gland, and advised excision.

On May 13th the operation was performed, and a good recovery rapidly followed. This lady is still well.

Description of **breast** *by Mr. Symonds.*—**Just be**neath the nipple was a cyst, one **inch in diameter,** with a thick dark wall, embedded in condensed breast tissue, containing many smaller cysts. Everywhere **through** the breast these cysts existed, and all **were filled with** a creamy grey fluid, which exuded as so **many beads** when the breast was squeezed. **The** galactophorous ducts were filled with this same material, which exuded from the nipple on compression. The whole organ had a uniformly whitish-grey colour, but nowhere was any solid growth visible. The out**line of** the breast seemed also normal, and there was no infiltration of, or alteration in, the surrounding **fat.**

Microscopical examination.—The fluid was not examined fresh, but from its resemblance to that seen in similar cases, and from its appearance in the hardened sections, it, no doubt, was composed of degenerated cells and fatty granules.

The appearances seen in the sections might be

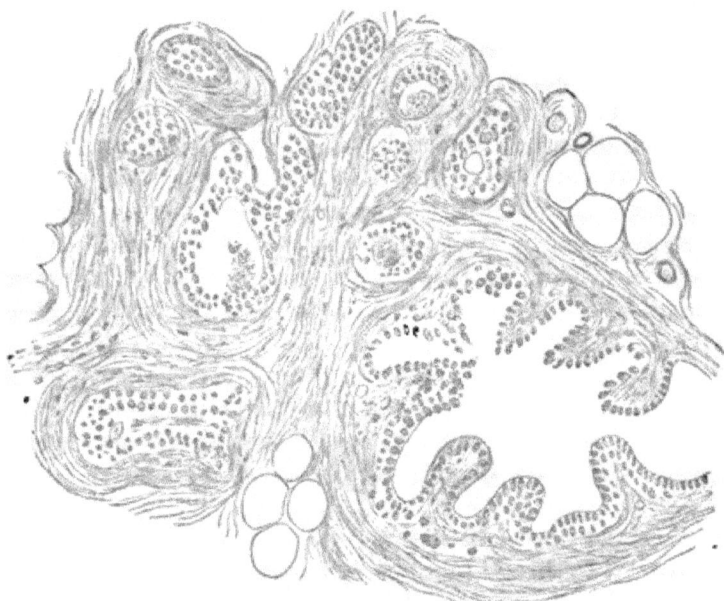

Fig. 7.—Breast undergoing Cystic Degeneration.   Case 1.

shortly described as showing the ordinary breast tissue containing many cysts (Fig. 7). There are groups of small tubes cut in various directions, some having a distinct lumen, many showing only a number of irregular cells. These groups are always surrounded by fibrous tissue containing fat vesicles, and represent the ordinary breast lobules. The adipose tissue is in some places tolerably abundant. Many of these tubes are much enlarged, having a wide lumen, occupied in

many instances by granular matter, and lined with
two or more rows of cells.   As the accumulation of
secretion increases, the central aperture enlarges until
a small cyst is formed visible to the naked eye. These
changes are all obvious in the woodcut below, and
occur in all parts of the breast.

Besides these cysts with a round lumen, there are
many apertures with a sinuous outline, and many
with villi or buds projecting towards the cavity.
These are all lined with two or more rows of cells,
the inner row being more or less columnar, and taking
the stain (logwood) more deeply than the outer or
irregular cells. The buds or projections are covered
by the same cells, and have vessels and fibrous tissue
as a basis. The appearances are shown in the illustration.

Again, the largest cysts have a wall composed of
lamellæ of fibrous tissue, are lined by large columnar
cells, and contain some of the granular material above
described. The cells form a definite layer much
altered from the regular columnar shape by mutual
compression ; usually the layer is two cells thick, but
there is no regularity in the arrangement. The fibrous
tissue seems no more in amount than that proper to
the breast, the cyst formation appearing to be the
only morbid change. From the description given
above, and from the appearances presented in the
drawing, these cysts all seem to form out of the mammary acini and ducts.

Case 2. *Cystic degeneration of breast which had
never secreted milk.*—Mrs. S., aged 34, the mother of
one child 17 years of age, consulted me in February,
1880, for some disease of her right breast. She had
suckled with her left breast, but not with her right,
as that gland had never secreted milk, although it in
all other ways appeared to be natural.

Three months before I saw her she accidentally

discovered a lump in the right breast, which had steadily increased. It was not the seat of pain, but only of uneasiness. When I saw her there were four or five nodules the size of nuts in the gland, and these, from their globular outline, I took to be cysts. The whole gland felt coarse to the hand, and knotty.

I punctured one of the nodules for diagnostic purposes, and let out some serum. The others I left. Six months later, when last seen, there was no change in the breast.

The cysts were doubtless due to cystic degeneration of the gland from involution changes.

GROUP II.—*Cystic tumours of the gland ; single or multiple ; of glandular, duct, or connective tissue formation without intracystic growths.*

That these cystic tumours of the mammary glands are not uncommon in practice must be fully recognised, although it must be acknowledged that they are too often not diagnosed until by some error they have been subjected to surgical operative treatment.

I take it few surgeons have been so fortunate as not to have removed a breast for a supposed cancer which turned out to be a cyst ; and fewer still who have not, as spectators of such a case, been well sprinkled with the fluid of a tense cyst accidentally opened by the hand of a surgeon in his attempt to cut into or excise what he regarded as a solid tumour.

Formation of cysts.—These cysts are doubtless developed in several ways. Some are unquestionably "*duct cysts*," that is, they are formed first by an obstructed and subsequently by a dilated irritated duct, the duct being more commonly a branch than a main duct, and occasionally a duct leading directly into an acinus of the gland. When the branch is a large one the cyst is likely to be single ;

when it is placed nearer the gland structure the cysts **are** more likely to be numerous. The cystic tumour in the former case appears as a globular, tense, more or less deeply placed growth ; and in the latter as an unequal enlargement of one of the lobules of the gland, **with a** more or less bossy outline and semi-fluctuating feel ; this latter symptom turning much upon the size the cysts attain and the depth of gland structure which covers them in.

The fluid contained by these cysts varies much. In **the** majority of cases it is clear and serous, in others it will be brown or slightly blood-stained, in a few cases it will be viscid and mucoid. When it is deeply blood-stained or sanguineous the surgeon should suspect that it is not a simple but a proliferating cyst in which some intracystic growth exists. **In** almost all cases the fluid will be albuminous ; at any rate, I have not yet met with an example in which **it** was otherwise. At times fluid may escape naturally or may be pressed from the nipple, and when this takes place the diagnosis of a duct or glandular cyst may with confidence be made. In half the cases of this affection no such symptom will, however, be found, but this fact need not diminish the value of the symptom for diagnostic purposes when it is present.

Others of the cysts have a "*connective tissue origin.*" That is, they originate outside the ducts of the gland, or the gland structure, and are formed by **the effusion** of fluid into the connective tissue which binds **the lobes and** lobules **of** the gland together. Such cases are usually single, but may be multiple. They are commonly of slow formation, and as a result have thick walls. They are always smooth on their **inner** surface, and as a rule contain serous fluid. At times the fluid is, however, mucoid or dark. When the cyst appears in the breast of a thin woman its

tense globular outline, unassociated with any of the
symptoms which are recognised as characterising the
existence of a cancerous or other tumour, should suggest its nature as well as its treatment, for the puncture
of the swelling with a fine aspirating needle will prove
much.

When the cyst shows itself in the breast of a fat
subject the diagnosis must be difficult, but its true
nature should ever be suspected, even when all other
symptoms of more solid growths have by the lapse of
time failed to manifest themselves. The more breast
tissue there is to surround a cyst or cystic tumour, the
greater the difficulty of diagnosis. The longer the
tumour has existed without the manifestation of the
well-recognised symptoms of cancer or other growth,
the stronger the probability of its being cystic. In
all doubtful cases, however, the fine exploring needle
of an aspirating syringe will avail much for diagnostic
purposes. In but few of these cases can the surgeon
without exploration diagnose between the simple cyst
and the cyst with proliferating intracystic growth.
Simple cysts of the breast exist, without in many cases
ever developing into anything else, or becoming the seat
of solid growth. With the majority of cystic tumours
this, however, is not the case, for what may seem to
be the most simple or innocuous cyst may be expected,
if not cured, to become the seat of some proliferating
intracystic growth, which will be adenomatous, sarcomatous, or carcinomatous, according to the tendency
of the tissue to form, and of the individual to develop,
either special variety.

The mere presence of a cyst in the mammary gland
must be accepted as evidence of some unnatural local
irritation, inflammatory or otherwise, whether that
irritation has originated in, or become localised about
a gland duct, lobule of gland, or connective tissue.
If the irritation is subdued or subsides, a cure, or at

any rate freedom from fresh developments of the local
trouble may be looked for; but should the local irritation continue or intensify, fresh developments must be
expected, and such of necessity will assume the form of
a new growth either of the glandular, epithelial, or connective tissue type, and lead on to the formation of an
adenomatous, epitheliomatous, or sarcomatous cystic
growth, to which attention will now be drawn. In
rare cases the cyst will suppurate and undergo a spontaneous cure. I propose to illustrate the clinical
history of this group of cystic disease by means of
cases, each case telling its own tale.

Case 3. *Duct cyst in breast with retracted nipple;
albuminous fluid drawn off; cured.* (Reported by
Mr. H. Caddy.)—Susannah B., a healthy looking,
well-nourished woman, aged 24, was admitted into
Lydia ward, under Mr. Bryant's care, on the 9th
September, 1872, with a large hard substance in the
centre of her left breast. The nipple had quite
disappeared, from retraction, but the swelling seemed
to point at the surface about an inch above where the
nipple ought to be. There was no glandular enlargement in the axilla or above the clavicle.

The patient's family history had been good, and
she herself had been healthy up to thirteen years old,
when her back began to grow out both backwards and
to the right side. She had no remembrance of any
injury, and a year later she was employed in turning
the mangle.

About four years ago her tonsils were cut, and
two years later she felt slight pain in the upper
part of her left breast, which became more severe
when she raised her arm. One month before admission she noticed a lump about the size of a
filbert, which was movable; a fortnight later there
was a discharge from the nipple. She then sought
advice from a medical man, who told her to bathe

R—25

it with warm water; this benefited her at first, but the lump grew, and the discharge from the nipple increased and became mixed with blood. She then came into the hospital.

On admission. A very tense globular swelling, the size of an orange, was present in the centre of the breast. The nipple was flattened out and somewhat retracted; it discharged a yellow glairy fluid, which could be increased on pressing the tumour. The diagnosis of cystic disease of the breast was made.

September 12th. A small cannula was introduced into the cyst, and about four ounces of a pale yellow fluid were drawn off. A drainage tube was introduced into the cyst.

16th. The breast discharged a little for a few days, but it ceased on the 19th, and on the 21st she left the hospital well.

Two years later the patient reported herself as still well.

Case 4. *Cyst in breast, which disappeared after tapping.*—Eliza T., aged 43, a married, childless woman, came to me February 12th, 1886, with coarse, indurated breasts, and a tense, globular tumour, the size of a walnut, in the left gland, which seemed to be a cyst; there was no discharge from the nipple. For diagnostic purposes this cyst was punctured, and clear serous albuminous fluid drawn off. By April 16th the swelling had gone. Six months later no return had taken place.

Case 5. *Duct cyst in breast, which spontaneously suppurated; was opened and cured.*—Eliza H., aged 40, a married, childless woman, came under my care in April, 1870, with a tumour in her right breast, which had been coming for five years and was increasing slowly. She had had serous discharge from the nipple for two years. The axillary glands were not enlarged. At present there is a free discharge of a clear fluid from the nipple, which runs **without,** but can be increased by pressure. At times the fluid is blood-stained. There is a tumour the **size** of **an** orange at the lower part of the gland. **An** incision was made into the cyst at its lower part, and some ounces of pus were evacuated. A good re- **covery** followed.

It is probable that the spontaneous suppuration of a simple duct cyst of the mamma is rare. I have not seen any other example than the one recorded, in which it was clear that this took place.

In Prep. 4753 of the College of Surgeons there is, however, a cyst of this kind, in which at one **spot** the inner surface of the cyst appears thinly covered with lymph, demonstrative of an inflamma- tory action. In this case, as in the one recorded in this paper, the nipple was retracted from traction **on** its ducts.

The following interesting case has been recorded by Mr. Birkett (Guy's Hosp. Reports, 1872-3), and illustrates the same point.

Case 6. *Duct cyst of the breast, following a contusion; retracted nipple; twelve years later spon- taneous suppuration and recovery.*—A delicate single lady, aged 40, consulted Mr. Birkett for a bloody dis- charge from the nipple, which she had observed for **nine or** ten years. She attributed it to a blow. The nipple was retracted. When seen a small, fluctuat- **ing,** painless swelling was present near the nipple, **with induration** of the gland tissue around it. By compressing the tumour, serum escaped from the nipple. Four years later, when but little change had taken place in the tumour, the discharge from the nipple ceased and the tumour inflamed. Suppura- **tion** soon followed, and the abscess burst, when more than ten ounces of pus escaped, and a piece of solid substance as long as the little finger escaped. This was the statement of the patient. The discharge, which continued **several** days, diminished ; the hole, which looked very deep at first, contracted and filled up, and at last became healthy. In fact, adds **Mr.** Birkett, the disease was cured by the efforts of nature alone, and has never reappeared in the eighteen **years which have** since elapsed.

Case 7. *Cystic tumour of breast; cured by* **an** *incision into its cavity, and drainage.* (Reported by Mr. Duckworth.)—Louisa N., a healthy looking **married woman,** aged 45, was admitted into Lydia ward, Guy's Hospital, on May 1st, 1879, under Mr. Bryant's care. Her paternal grandmother had cancer of the breast. **About three** months ago she felt some pain in the region of the left breast, but noticed no lump for a month, when she felt one about the size of a walnut on the outer side of the gland. **It was** freely movable, and when she slept on her right side would slip towards the inner side. **It was occasion-**ally the seat of a gnawing or shooting pain. She has had one stillborn child, and never suckled.

On admission. There **is** a hard, irregular, ovoid, lobulated swelling about the size of an egg, on the outer side of the left breast. It is very freely movable, and the skin over it is not infiltrated. The tumour seems, however, to be slightly attached by one end towards the nipple, which is normal.

Operation.—May 6th, under chloroform, an exploratory incision was made about two inches in length on the outer side of the breast into the tumour, which was found to be a cyst. When opened about an ounce of clear fluid escaped. The **cavity,** which **was** smooth, was washed with iodine and water, after which a drainage tube was put in and the wound **dressed** with lint soaked in terebene oil.

8th. No discharge, wound healthy, drainage tube removed. Temp. 97·8°.

12th.   Tinct. ferri perch. ℥x, tinct. calumb. ℥xx, aq. ʒj, t. d. s.   Temp. normal.

20th.   The wound has nearly healed.

24th.   Discharged cured.   Some months later this patient was well.

Case 8.   *Cyst in breast, simulating carcinoma; exploratory incision; plugging of cyst; recovery.*— Martha J., aged 50, a married woman, came under my care on January 9th, 1886, with a tumour the size of an orange, occupying the centre and axillary half of the left breast, retracted nipple, and healthy skin over the tumour.   No axillary glands found enlarged. She first noticed the lump five years previously, which had steadily and almost painlessly increased.   It was the seat of an occasional shooting pain.   The tumour and breast were one ; it was hard and inelastic ; no nipple discharge. An exploratory puncture or incision was determined upon.   On January 19th, when the woman was under the influence of an anæsthetic, fluctuation was detected ; an incision was then made into the tumour in a line radiating from the nipple, and a cyst opened ; three ounces of a dark-coloured fluid, containing cholesterine, was evacuated.   No intra-cystic growth was found, and the lining membrane of the cyst was smooth.   The cavity exposed was then plugged with iodoform gauze dipped in terebene oil to excite suppuration, and the case did well.

Case 9.   *Cyst in breast treated by incision; doubtful thickening remaining; well.*   (Reported by Mr. Duckworth.)—Harriet S., a single woman, aged 35, was admitted into Lydia ward on May 21st, 1879, under Mr. Bryant's care.   Six years ago she had an abscess in her neck. About four months ago the patient experienced, when at work, a dragging pain in her chest and shoulder, and found upon examination a small hard lump, about twice the size of a pea, in her right breast.   It was freely movable, and

occasionally gave pain at the back of her shoulder.
She has no recollection of having received an injury.

On admission. To the left of the nipple of the
right breast there is a hard, irregular, lobulated lump
the size of a small orange, with a base occupying about
the area of a half-crown. It is freely movable, and
does not infiltrate the nipple or breast. The axillary
glands are slightly enlarged.

May 21st. She is menstruating; her menses have
been irregular for some time. They generally last
ten days.

27th. Under chloroform, an incision was made
over the tumour, radiating from the nipple, and about
two and a half inches in length. Having cut into the
tumour, about an ounce of brownish fluid escaped;
the walls of the cyst were smooth. The parts were
washed out with iodine and water, and a drainage
tube introduced; the edges were then partially
brought together with four sutures.

28th. Temp. 98°, pulse 88.

29th. Temp. 98·2°. Two sutures taken out;
wound looks healthy.

June 1st. Temp. 98·9°, pulse 96. Wound closed
well, narrow line of granulations on surface. A lump
about size of a hazel nut felt below the incision.

24th. Patient left the hospital cured.

Case 10. *Cyst in right breast; treated by a free
incision; no intracystic disease; patient well four years
later.* (Reported by Mr. Gowan.)—Mary S. R., aged
40, was admitted on October 15th, 1883, into Lydia
ward under Mr. Bryant's care. Patient is a tall,
well-formed, and well-nourished woman. She is a
widow. Has never borne a child, and has never been
injured in the breast. Two months ago she first
noticed a lump in her right breast about the size of
a pigeon's egg, a little to the right of the nipple. It
has grown somewhat rapidly since.

On admission. The right breast does not appear much larger than the left. **Both are** well developed and firm. The right nipple is slightly tinged with brown, but not retracted, nor is the skin dimpled. A tumour, about the size of a large orange, occupies the upper and outer quadrant of the right breast ; **it is** globular and smooth, and indistinctly fluctuates. Neither the skin nor adjacent muscles are implicated apparently.

October 16th. Under chloroform the tumour was punctured with a scalpel, when about one and a half ounces of blood-stained serum gushed out. The incision was enlarged, and the cyst walls were found to be free from growth, smooth and thin. The wound was plugged with iodoform gauze, and dressed with gauze and flannel bandages ; the arm was fastened to the side.

November 19th. Left hospital ; there is still an open granulating wound $1\frac{1}{8}$ inches long by $\frac{3}{4}$ inch. Patient's general health is good.

Three years later this patient was quite well.

Case 11. *Deeply placed cyst in sternal lobe of right breast, simulating adenoma in gland ; exploratory incision ; plug.* (Reported by Mr. C. Lloyd Jones.)— Margaret V., a domestic servant, aged 37, was admitted **into** Lydia ward on February 23rd, 1877, under Mr. Bryant's care.

She stated that her father died at seventy years **of age of** some internal tumour, and her mother at **thirty** of hepatic disease. Her brother and sisters **were all** healthy. She was unmarried, and always enjoyed good health. She fancied that she had become a little thinner during the past winter, and lately her menstrual intervals had become three weeks instead of a month. Three weeks before admission she happened to place her hand on her right breast, **when** she noticed a lump **there**. She felt no pain in it, and it had not increased in size since it was first noticed.

When admitted, the hand pressed flat against the breast at once detected, at the upper part of the right breast, a tumour which was nearly circular in shape, about two and a half inches in diameter, and flattened. It felt hard, and its surface was distinctly lobulated. The tumour was movable with the breast, the skin was nowhere puckered nor adherent, and there was no pain on manipulation. The nipple was normal. No enlarged glands in axilla or elsewhere.

The patient could assign no cause for the appearance of the tumour, and she stated that she was, as she appeared to be, in very good health.

March 9th. Under an anæsthetic an incision was made into the tumour, when blood-stained serum escaped ; the lining membrane of the cyst which occupied the posterior part of the lobe was smooth ; the cyst was plugged.

15th. Wound nearly well. The patient had not had one bad symptom or any rise of temperature.

Remarks.—In this case the true diagnosis was not at first made, nor was the true nature of the case suspected ; the tumour had none of the appearances of a cystic growth, since it was quite inelastic, and its surface was nodular and hard.

The cause of this error in diagnosis is to be explained by the position of the cyst, which was covered in by gland of quite three-quarters of an inch in thickness, the gland giving the nodular outline and yielding the firm feel.

At one time the question of infiltrating carcinoma was considered, on account of the growth so closely involving the gland and appearing as an infiltration, but this was dismissed in favour of the adenoid view, the outline of the tumour appearing more of the latter disease.

Had this case been punctured for diagnostic purposes its true nature would have been discovered.

Case 12. *Cyst of mamma mistaken for carcinoma, and excised ; cure.* (Reported by Mr. C. E. Perry.) —Anne F., a married woman, without children, aged 47, was admitted into Lydia ward, under Mr. Bryant's care, on January 18th, 1875, with a hard round swelling, about the size of a small tennis ball, in the upper and outer part of her left breast ; it was movable with the breast and without pain except on pressure. It was regarded as a carcinoma. It appeared that an aunt on her mother's side had died of cancer in the breast. She herself had had good health until about five months before admission, when she felt a swelling about the size of a small egg in her left breast ; it remained about the same, without pain, for three months, when, pain coming on, she sought advice, and had some embrocation, which she left off after a few days, as it brought out a rash. She then fomented it, and nothing else was done up to the date of her admission.

January 26th. Chloroform having been given, the breast and tumour were removed by two elliptical incisions. On cutting into the tumour it was found to be a simple cyst full of fluid.

The patient did well after the operation, and left the hospital convalescent.

The preparation is in the Guy's museum, No. 2290^{80}.

Case 13. *Sero-cystic disease of right breast; removal of lobe ; inflammation of the remainder, and recovery.*—Mrs. S., a childless, married woman, aged 50, was brought to me by Dr. Wallace, of Hackney, March 15th, 1875, with a tumour in the axillary lobe of her right breast, which had been steadily increasing for three months. The tumour was hard and lobulated, and I supposed the growth to be simple, and advised its excision.

March 21st. On excising the tumour I found it
to be made up of cysts, and in making the section
of the gland the whole breast was found to be full of
small cysts (involution cysts), many of which I punctured. I could not remove the whole gland, as I had
undertaken to remove the tumour alone.

During convalescence the remainder of the gland
inflamed and indurated, putting on the clinical
features of acute cancer. The induration, however,
subsequently entirely subsided, and the patient recovered.

In 1881 she was quite well.

The tumour on removal was made up of cysts of
various sizes, the largest being about the size of a
walnut. The cysts contained fluid of different kinds.
In some it was clear, in others blood-stained, whilst
in a few it was mucoid and greenish. There were no
intracystic growths. Some of the ducts were open,
and through these bristles could be passed.

The gland was clearly undergoing cystic degenerative changes, and in the lobe removed these changes
had gone on rapidly.

Case 14. *Sero-cystic disease of breast following
injury; excision of gland; patient well six years
later.*—Miss D., aged 34, patient of Dr. S. Tayleur
Gwynne, of Whitchurch, Salop.

In June, 1864, she had a blow upon her left breast,
which was followed by pain for two months, and then
the appearance of a swelling.

In January, 1865, when I saw her, the swelling
clearly occupied the sternal half of the gland, and was
tense, giving the idea of a cyst. General health good.
By December of the same year the tumour had
increased, and had become lobulated; it, moreover,
was more painful. Nipple and lymphatic glands
normal.

December 29th. Breast removed and found to be

a pure cystic disease of the gland. Cysts contained brown thin fluid, some clear serum. No intracystic growths.

Rapid recovery ensued, and six years later the patient was well.

Case 15. *Sero cystic disease of breast; excision of breast; patient well six years subsequently.*—Mrs. R., aged 40, the mother of seven children, was brought to me by Dr. Wilton, **of Sutton. Her** left breast **has** no nipple; it had been destroyed by old ulceration. The right breast is the seat of cystic disease, which involved the whole gland, and had been going on for months. The gland was very large, with an irregular outline, and clearly contained cysts of all **sizes.** The nipple and skin over the breast were normal.

January 29th, 1881. Breast removed, and **the** operation was followed by a **good recovery. The** gland was full of cysts of all sizes, **which contained a** great variety of fluids. No intracystic growths could be found.

January, 1882. Pregnant. Natural delivery. Left breast strapped. No nursing trouble.

1887. This patient is now well.

Case 16. *Sero-cystic disease of breast; excision of breast; patient well eleven years later.*—Mrs. S., aged 39, the mother of three children, all of which she had nursed without trouble, the youngest being fifteen years of age, was brought to me by Mr. Joseph Burton, of Blackheath, in 1876, with an irregular nodular **tumour** in her left breast, which had been coming on several months. It was apparently cystic, and involved the whole gland. The skin over the gland was normal. **Her mother had** had cancer **of** her breast, and the sister some internal tumour.

February 16th, 1876. Breast removed. On **section a** large cyst without intracystic growth, containing **fluid** as **dark** as ink, was found, with many smaller cysts containing fluid of different characters. In fact, the whole breast was full of cysts of different sizes. A good recovery ensued; and in 1887, eleven · years later, the patient was well.

GROUP III.—*Cystic tumours of the breast of whatever kind in which papillomatous, adenomatous, sarcomatous, or carcinomatous intracystic growths are present.*

If, clinically, simple cystic tumours of the breast are fairly common, pathologically they must be pronounced to be comparatively rare, for it is, I think, indisputable that in the majority of cystic tumours of the breast, of whatever character, of duct cysts, gland cysts, or connective tissue cysts, some solid element can usually be found proliferating from their lining walls, either in the shape of a small sessile or pedunculated outgrowth, or of a solid tumour composed of glandular, epithelial, or connective tissue elements; the nature of the growth being determined by the seat and persistency of the local source of irritation which originated the formation of the cyst. When the cyst is single the growth will of necessity involve it alone. When multiple, some of the cysts may be simple or non-proliferating, while others are proliferating, the association of the two kinds in the same gland being not uncommon. The surgeon consequently meets with cystic tumours of the breast, in which in one case *papillomatous growths* are present; the papilloma being a simple outgrowth of the cyst wall, in the same way as the same growth may spring from the nipple or other structure; whilst in another case the cyst may have an *adenoid growth* hanging or springing from its walls, the growth appearing either as a pedunculated intracystic growth washed with the cyst contents, or the cyst may be filled with sarcomatous or adenoid tissue either of a loose or more

solid structure; the tumour in the latter case, when
the cyst is full, losing much of its cystic character and
approaching the more solid kind, whilst the intermediate
conditions between these two extremes suggest and
give support to the view which some pathologists
entertain, that these solid adeno-sarcomatous or fibromatous tumours **as a** rule originate in cysts, and only
differ in the degree in which the cyst cavity is filled.

In many cases, doubtless, what appears to be an
intracystic growth is nothing more than a growth
originating in the connective or glandular structure
outside the cyst, but projecting into it, and having its
surface bathed with fluid, the fluid being either the
secretion of the cyst wall or an exudation from **the**
growth itself into the cyst cavity. The same remarks are also applicable to sarcomatous and carcinomatous tumours. Without doubt a large number of
tumours of the breast begin as cystic, and pass **on to**
become sarcomatous and carcinomatous; **the con-**
nective tissue element in the former case, and the
epithelial element in the latter, so increasing and
filling what was at one time simply a cyst or cysts, as
clinically to form what is now known as a cystic
sarcomatous, cystic adenomatous, or cystic carcinomatous tumour. These varieties in the cystic disease
maintain the same clinical features as characterise
the more solid kinds; the symptoms of the solid and
the cystic being the same in each, with the addition
in the latter of the cyst element. The diagnosis of a
cystic adenoma, cystic adeno-fibroma, sarcoma, or carcinoma is consequently to be determined by the same
points as are known to characterise a solid adeno-
fibroma, adenoma, adeno-sarcoma, or carcinoma, with
the addition of such clinical symptoms as are clearly
referable to the existence of cysts.

The following cases will best illustrate the clinical
features of the disease.

Case 17. *Cystic (duct) adenoma of breast; excision of lobe.*—Mrs. D., aged 45, the mother of four children, the youngest of whom was eight, and who had only nursed her first child, consulted me in April, 1876, for a tumour in her right breast, which had been discovered six weeks, and had steadily increased. I took it to be a cyst. The swelling was central and very cystic to the feel. The nipple projected more on the right than the left side, and discharged blood-stained serum. The skin over the gland was healthy and the lymph glands were natural. I advised puncture or incision, and excision if the cyst was found to contain growth. The late Mr. Jardine Murray, of Brighton, operated in May, 1876, and found a pedunculated intracystic growth, which, on examination, I proved to be an adenoma. He then excised the affected lobe, and a good result followed.

Five years later this lady was well.

Case 18. *Case of cystic adeno-fibroma of the breast, in a woman, aged 71 ; excision of gland ; cure.*—Catherine K., aged 71, a healthy looking woman, the mother of three children, was admitted into Guy's Hospital, under the care of Mr. Bryant, on August 7th, 1865. She had always enjoyed good health, and had been able to suckle her children. About last Christmas, some time before admission, she accidentally discovered a lump, the size of an egg, in the outer side of her right breast ; it was painless, and grew very slowly for three months, when it suddenly began to increase rapidly in size, and to cause pain. She applied to Mr. Bryant for relief, and remained under his care till she was admitted. On her admission the right breast presented a large tumour, the size of a cocoanut, closely connected with, if not in, the breast. It was irregular in its outline, and evidently in parts made up of cysts, for, in its projecting portions, distinct fluctuation was clearly felt. There was, however, much solid matter. The tumour was quite movable, and the skin over it was only stretched ; the axillary glands were also healthy ; the nipple was natural.

Nothing but excision promising to be **of any use,** the operation was performed on August 30th, and a rapid recovery **took** place, the old woman leaving the hospital **in one** month perfectly well.

On examining **the** tumour it was found **to be made** up of a firm solid material, which **contained several** large cysts. These cysts held a blood-stained glairy fluid, and in parts the solid growth seemed to threaten to degenerate and break **up.**

The tumour measured seven inches **by** six ; it **was** very firm in its consistence, and to **the** eye appeared **of** a sarcomatous nature ; it was tough, and was with **difficulty** broken down ; in parts, however, it had more the aspect of the looser kind **of** adenoid tumours.

By the microscope, the opinion formed by the naked eye examination was confirmed, for the structure generally was an admirable specimen of the more fibrous kind of adenoid tumour ; tubes were here and there visible, and also well-developed cell structures, **as a drawing** by Dr. Moxon indicates.*

As an example of an adeno-fibroma **in an old** woman, the case must **be regarded with great** interest.

Case 19. *Cystic sarcomatous disease of the breast of fifty years' growth in a woman seventy-three years of age.*—E. C., a healthy looking old woman, 73 years of age, came under my care on October 5, 1862, with an enormous tumour, the circumference of a soup plate, in her left breast. She was a married woman, the mother of one child, who was fifty years old. During **her** pregnancy with that child, she observed a small lump in her left breast, which was movable. This

* *See* Path. Trans., vol. xvii. p. 283.

gradually increased in size, but in such a painless way, that she kept her trouble to herself; about three months before she came to me the tumour burst and discharged much blood-stained serous fluid; it was on this account alone that she sought advice. When I saw her a large bossy tumour occupied the position of the left breast; it was fluctuating in parts, and the skin over it was healthy, though much thinned. The nipple was flattened out over it. At one part the skin had ruptured, and the rent appeared as a fissure, with a thin but healthy margin, and through this opening sprouted a large fungating growth. The absorbent glands were natural, and the patient's health was good. No operation was sanctioned.

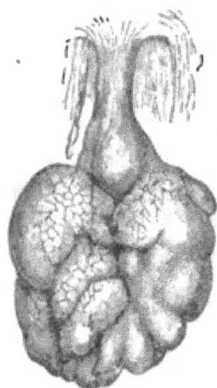

Fig. 8. — Pendulous Intracystic Adenoid Growth. Case 20.

Case 20. *Cystic (duct) adenoma of breast, excision of gland. Patient well six years later.*—Mrs. T., aged 54, the mother of several children, all of whom she had nursed without trouble, consulted me in October, 1880, for a tumour of her breast which she had noticed for about a year. The appearance of the tumour had been preceded for months by the discharge of a serous fluid from the nipple. The tumour was about the size of a tennis ball, hard and globular; pressure upon it caused blood-stained serum to flow from the nipple. The skin over the swelling was natural, and the nipple was not changed. The diagnosis of a duct cyst was made, and an incision advised into the tumour with the view to an excision should an intracystic growth be found. This was performed on October 23rd, 1880, and the breast removed, as a pedunculated intracystic growth was found in

the cyst (Fig. 8).   A good recovery ensued, and
the patient was well six years later.

The growth was examined by Mr. Symonds, who
reports :

The growth was one and a quarter inches in
diameter, and was attached by a small pedicle to the

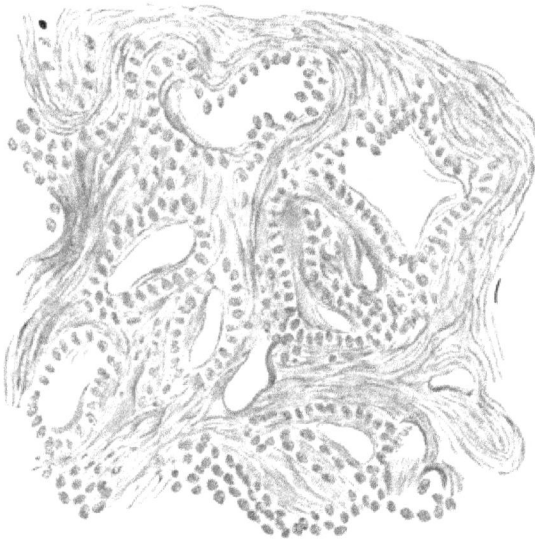

Fig. 9.—Microscopical Appearance of Breast Adenoma.   Case 20.

smooth wall of the cyst in which it was enveloped.
It had a deep purple-red colour, was soft, and separated easily from its attachment.   Round the cyst
and in its wall were many dilated ducts, some of large
size, but no direct communication could be traced.
The cyst appeared to be produced by dilatation of a
duct.   Nowhere in the breast was there any other
growth.

The tumour had a lobulated appearance, as exhibited in Fig. 8, and its pedicle was small ;
microscopically the growth showed all the characters

s—25

of a pure adenoma as seen in drawing (Fig. 9).
It is composed of large spaces of various shapes,
lined by closely set cells. These have large nuclei
and one or more nucleoli. The supporting framework
is a very delicate fibrous tissue, containing cells of
various shapes, besides blood-vessels. From the walls
of many acini conical projections grow, which often
blend with the opposite wall.

The tumour therefore appears to be a pure adenoma growing from the wall of a duct. The small
amount of fibrous tissue, and the normal characters
of the breast tissue immediately round the cyst,
negative the view that it is a fibroma budding into a
dilated duct. This growth is identical with that
called now a "duct papilloma."

Case 21. *Cystic duct tumour with pedunculated
adenocele of breast; excision; patient well three years
later.* (Reported by Mr. W. T. Crew.)—Susannah
W., a single woman, aged 43, was admitted under
Mr. Bryant's care, into Lydia ward, on February 2nd,
1876, with a globular tumour involving the whole of
her right mamma, and extending into the axilla, on
the inner side of which there were little hard nodules
like peas. The tumour within the breast gland was
freely movable under the skin and over the pectoral
muscles; the veins were distended over the tumour.
Fluctuation was very distinct over the greater part
of the swelling; there was no discharge from the
nipple, and the lymphatic glands were unaffected.

In 1872 she first noticed an exudation from her
nipple like glycerine; soon after, a lump the size of a
walnut appeared in her breast without pain, and the
swelling increased. The breast would sometimes feel
tense, but become relieved as soon as the discharge
took place. For six weeks there had been no discharge.

February 3rd. The patient having a little cough

the operation was deferred until the 8th, when chloroform having been administered, an elliptical incision was made, and the whole tumour, with nipple and a portion of the skin, were dissected out and removed.

The tumour was mainly composed of a cyst which contained about half a pint of mucoid fluid, and was lined by a thin, delicate, white membrane ; attached

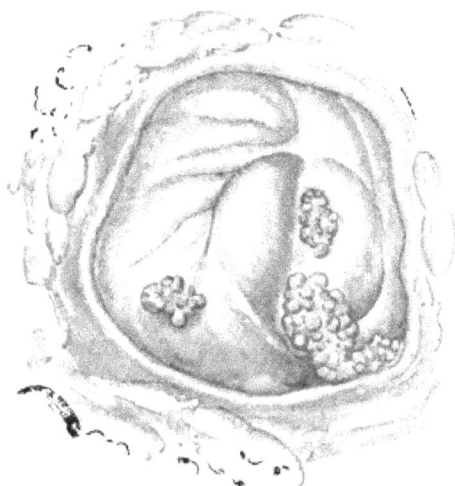

Fig. 10.—Cystic Duct Tumour with Pedunculated Intracystic Growths.
Case 21.

to its walls were several lobulated pedunculated tumours as seen in woodcut (Fig. 10).   The growth examined showed a well-marked papillomatous structure.   The processes are covered by long columnar cells, and often divide.   Circular or elongated apertures lined by the same cells are also numerous.   There are also small cysts with definite fibrous walls and colloid contents.

The growth closely resembles that in Mrs. T.'s case, No. 20, and suggests that the large cyst in this instance is also derived from a duct.   The small cysts in this growth have been formed, I suppose, by the

fusion of the processes in the manner described by
Wilson Fox. Several small adeno-fibromatous tumours were also found on making sections of the gland
in other parts of the breast. The arteries were twisted,
wire sutures inserted, pressure was exerted by pads of
lint, and the whole wound drawn together by strips
of strapping.

21st. The wound has healed by first intention.

Three years later the patient was well.

Case 22. *Cystic duct adenoma; excision of
tumour; cure.*—Harriet W., aged 47, the mother of
thirteen children, youngest six years, all of whom she
had suckled, was brought to me on January 14th,
1867, with a globular tumour, apparently cystic, in
the upper part of her left breast, which she had observed six months. At first she had a clear discharge
from the nipple, but she had not had any lately. The
skin over the tumour is apparently bound to the
growth. Axillary glands sound.

September 15th. Serous discharge from nipple
very free ; increased by pressure. Tumour large.

March, 1868. Tumour removed and found to be
composed of a cyst, with intracystic pedunculated
growth, which was pronounced by Dr. Moxon, after
microscopical examination, to be adenoid.

Case 23. *Duct cyst of the breast, following a
contusion; intracystic fibro-cellular growths; rupture
of cyst seven years later ; removal of breast ; recovery ;
patient well twelve years later.*—In 1860 Mr. Birkett
saw a lady, aged 45, who for seven years had had some
tumour in her right breast. In 1853 she received a
blow in the part, which was followed by a flow of
blood from the nipple. Two months later an abscess
formed, and was opened, a cupful of matter escaping.
The abscess healed, but some hardness remained, and
occasionally a bloody fluid oozed from the nipple. At
times half a pint would flow daily.

Subsequently a soft tumour formed and grew, and the nipple retracted ; and two weeks before consulting Mr. Birkett the skin over the tumour ulcerated, and a quantity of blood and fluid escaped. When seen, a soft, vascular, flocculent growth, **very** like **everted** mucous membrane, projected slightly from an opening in the integument, with which, however, it was *not* connected, and from which serous discharge flowed. There was no pain. The nipple, which had **been in**verted, was now everted. The axillary lymphatic glands were tender, but not infiltrated. Her general health was good. The whole breast was removed, and a rapid recovery followed. The patient twelve years later was well.

A section through the disease exposed a cyst, with **firm** fibrous walls, embedded in adipose tissue. **Three** pedunculated masses of soft, vascular new growth were hanging from the cyst wall. From the nipple a bristle was passable along a duct into the cyst. The intracystic growth was composed of fibre tissue, nucleated bodies **of** variable shapes, and a reticular struma with an arrangement like papillæ or villi. The true gland tissue was nowhere visible.

Case 24. *Cystic duct sarcoma of the breast which never secreted milk.*—Mrs. G., aged 45, the mother of one child, aged 17, consulted me in February, 1881. She had suckled with her right breast, but not with her left because she had no milk in it, and the nipple was retracted. For six years she had had a discharge **of a** clear fluid from the retracted nipple of th left breast, and which at the catamenial periods was increased ; the fluid at times was like " treacle water." Three months ago this flow stopped and a lump appeared.

A tumour the size of a walnut exists in the centre **of** the breast beneath the nipple. There is no dis**charge** from the nipple. I advised incision and excision if necessary. **Dr.** Godfrey, of Balham, subsequently tapped the cyst and discovered a growth, the excision of which was advised, but not acceded **to.**

Case **25.** *Cystic sarcoma of the breast fourteen years ; removed ; well seven years later.*—Mrs. B., aged 42, of Bottesford, Nottingham, consulted me in 1877 **for a tumour in** her left breast of fourteen years' duration. **She was the** mother of four children, the **youngest being four** years old, all of whom she had **suckled. At the birth of the last** child, four years **ago, the** tumour was no **larger than** an egg. For the **last six** months it had grown rapidly, and is now as large **as a fist. It is** clearly cystic from its nodular shape and fluctuating feel. There is no discharge from the nipple. No lymphatic glands are enlarged. Excision was advised and performed, and a good recovery followed. In 1884, seven years after the operation, the patient was well. Her maternal grandmother had died from cancer of the tongue, and her paternal grandmother from cancer of the breast.

Case 26. *Cystic sarcoma of the breast ; retracted nipple ; excision ; well eight years later.*—Mrs. S., aged **50, the mother of** nine children, one only of whom she suckled without difficulty, consulted me in March, **1876,** for a tumour in her breast, which had **been steadily** increasing for two years. Eight of her children **had died of** phthisis **at ages** varying from fifteen to twenty-seven. When **seen the** tumour was the size of a fist, smooth, slightly lobulated, globular, and elastic ; it appeared to be **cystic.** The nipple did not discharge, but was retracted. The axillary glands were free.

April, 1876. **It was** excised by Dr. **Sams, of** Blackheath, **who sent me** the **specimen.**

April 11th, 1876. **Dr.** Goodhart reported : " The tumour is spindle cell sarcoma or recurrent fibroid, as I expected from its lobulated appearance and manner

of growth. They are not at all uncommon in the
breast, and are often called adenoid growths from
their very similar naked eye appearance. It will
probably return in the cicatrix, though glands are less
likely to be involved in the axilla."

In 1886 this lady was still well.

Case 27. *Cystic sarcoma of breast; rupture of
cyst; hæmorrhage; excision; recovery.*—Mrs. S., aged
71, consulted me in January, 1873, for a tumour of
the breast of twelve years' growth. It was then the
size of a child's head. One cyst had ruptured, and
an intracystic growth sprouted from its centre which
bled freely. On this account an operation was advised, and performed on February 26th, 1873. A good
recovery ensued. The disease was on examination
found to be one of cystic disease, with intracystic sarcomatous spindle-celled growths. This lady died,
more than three years after the operation, in September, 1876, from acute bronchitis.

Case 28. *Cystic duct adeno-sarcoma of the breast
of several pounds weight, as indicated by free discharge
from the nipple, into which hæmorrhage had taken
place; excision; recovery.*—Mary A. W., a healthy
looking single woman, aged 20, came under my care
March 26th, 1861, with a tumour in her right breast
which had been growing for three years and a half.
It had commenced as a swelling the size of an egg,
which was situated in the centre of her breast, and
had steadily increased without pain. When coming
under observation the tumour was of the size of an
orange, globular and tense. I diagnosed it to be of
cystic origin. She disappeared from view for eighteen
months, and when she reappeared the tumour measured fourteen inches in diameter and twenty-two
inches in circumference. It was globular, with the
nipple in the centre; quite movable with the breast,
and uniformly elastic; fluctuation was readily to be

detected in it. The skin was much distended over
the tumour, and on its outer side it was red and inflamed from distension. The breast gland could not be
separated from the tumour.

The tumour during the interval mentioned above
had grown slowly up to six weeks, when its increase
became rapid. For three weeks there had been **a**
free discharge of a bloody fluid from the nipple, which
could be readily squeezed out by pressure upon the
tumour.

On **March 15th** I tapped the tumour, **and** let out
a quantity of bloody serum and broken-down blood;
to allow of its more ready evacuation the opening **was**
enlarged, when the finger, without force, easily broke
up a large portion of the growth. Two and a half
pounds of this material were thus taken away. The
largest portion of the tumour was, however, of a
more solid nature, which necessitated the removal
of **the** whole gland. This was done, and **a good**
recovery followed.

A section of the tumour through its centre and
the nipple showed a fine example of cystic sarcoma of
the breast. At **its** lower part were many beautiful
examples of intracystic adeno-sarcomatous growths,
which turned out of their cyst walls. About the
centre of the tumour were smaller growths, infiltrated
with blood, **and** breaking down; and in the upper
part there was little else than extravasated blood and
clot. Microscopically, **all** the elements of sarcomatous tumours were present, spindle-cells being
abundant; there were, too, some glandular elements
such as are found in adenoma.

Case 29. *Cystic sarcoma of the* **breast;** *rupture
of cyst; operation refused.*—Clara W., aged 43, **the**
mother of three children, the youngest being fourteen
years old, consulted me on March 7th, 1867, for a
tumour the size of an orange in her left breast, which

had been growing thirteen months. It was globular, semi-elastic, and nodular. The skin, nipple, and axillary glands were sound.

April 7th. Tumour much larger; its bossy surface is lost, and the growth seems to be one large cyst. Cyst tapped, and about one ounce of blood-stained fluid drawn off.

June 2nd. Tumour size of cocoanut.

July 8th. Tumour burst and discharged fluid largely mixed with blood. Operation refused. Intracystic growths visible; later on some of the growths projected through the rupture in the cyst wall.

July 16th, 1868. Patient sinking from asthenia and bleeding from the growth.

Case 30. *Cystic sarcoma of breast; rupture of cyst; operation rejected.*—Mrs. R., aged 63, married at age of twenty-one and had five children, the youngest aged thirty. Nipple in left breast always retracted. Thirty years ago, independently of nursing, she had a tumour the size of a fist in her left breast, which disappeared after one year. She remained well for twenty-eight years, up to one year ago, when the swelling reappeared in the same breast, and this has gradually grown.

At present (July 9th, 1873) a tumour the size of an egg exists in the upper part of the breast, which seems to be a cyst; this is adherent to the skin. Excision of the tumour was advised but not accepted.

January 19th, 1874. The cyst, which has much increased in size, has ruptured and discharged its contents, a sarcomatous growth protruding from the skin opening. No operation was sanctioned.

Case 31. *Cystic sarcoma of breast; suppuration of a cyst followed by rupture of cyst; no operation.*—Mrs. C., aged 43, of Hull, the mother of five children, her youngest child being now fourteen years of age.

After giving up nursing there was discharge from her nipple of a clear greenish or blood-stained fluid. Six months ago this discharge ceased, and the breast enlarged, inflamed, and suppurated. It was opened at that time by Mr. W. H. Rudd (April, 1883), and the abscess healed, leaving a lump. This increased in spite of treatment, and is now the size of a large egg. It appears to be (February, 1884) cystic. Incision into the cyst was advised, and excision of the lobe advised if growth was found. Nothing was, however, done, but the cyst burst and discharged bloody fluid, and discharge has been going on up to the present (October 12th, 1884).

Case 32. *Cystic sarcoma.*—Mrs. S., aged 70, has had three children, which she never suckled. In 1875 she consulted me for a tumour in her right breast, which she had had four years; when seen by me it was the size of a cocoanut, clearly cystic, and nodular. No lymphatic glands were enlarged. Skin normal. Operation not advised on account of age.

Case 33. *Cystic fibroma of right breast, with cystic degeneration of the gland; excision of tumour; cured.* (Reported by Mr. Metzgar.)—Rebecca B., aged 40, was admitted into Lydia ward on June 21st, 1884, under Mr. Bryant's care. Patient's father and two brothers died of phthisis, and she frequently suffers from bronchitis and neuralgia. She is married, and has one son alive, and healthy. Six weeks ago she noticed an uncomfortable swelling in her right breast, which appeared as a prominent tumour in the upper half of the gland. This tumour has not increased much since then, but it has seemingly got softer.

On admission. The tumour is hard and nodulated, firmly embedded in the gland, but freely movable with it upon the pectoral muscles. The skin over it is normal. The axillary glands are not enlarged. Nipple natural. Urine normal.

June 24th. Under chloroform an oblique exploratory incision was made into the tumour. The tumour, **when** incised, appeared granular and fibrous, a fibroma apparently growing from or into a cyst ; the breast itself had undergone cystic degeneration. The tumour was then freely separated from the surrounding breast tissue and removed. In so doing black cysts were seen spread over it. The vessels having been twisted, and capillary hæmorrhage stopped by hot iodine sponges, three silk sutures were put in, and **one and a** half inches of drainage tube were inserted into the upper and left end of the incision, the lower or right half being well padded with lint and a sponge. The arm was strapped to the side.

June 25th. Drainage tube removed.

30th. Stitches removed. Lips of wound drawn **together** with waterproof strapping.

July 11th. There has been good primary union. There is very little discharge. A plug of terebene lint into the lower and right end of **wound**.

26th. Went out, to come up from time to time to be seen. This patient was subsequently quite well.

The tumour when removed proved to be a fibroma growing from the walls of a cyst. The gland itself was full of small cysts, the result of degeneration.

The temperature only on the occasions after the operation reached 100°.

Case 34. *Recurrent mammary cystic sarcoma ; excision ; well.* (Reported by Mr. H. H. Wright.)— **Eliza M.,** aged 36, was admitted into Lydia ward on July 14th, 1880, under Mr. Bryant's care. Patient is a married woman, and the mother of eight children. Three years ago she was operated upon for tumour of the left breast ; the tumour was removed, but the breast was left. She has since been able to suckle **a** **child at** the same breast. A year ago she noticed a small lump above the nipple which has steadily

increased. On admission there was a large lobulated
tumour involving the whole of the left breast, but
divided into two chief parts. One of these occupies
the central and upper part of the gland ; the other,
which is smaller, is placed on the axillary side. **The
whole tumour is above the level of the old cicatrix.**
It is freely movable, partly hard and partly elastic to
the touch. One point above and to the inner side of
the nipple is inflamed and painful.

July 20th. Under an anæsthetic the whole breast
was excised.

The patient suffered a little from conjunctivitis
after the operation, but otherwise did well, and on
August 16th went out, the wound having healed.

Description of tumour.—The tumour was everywhere encapsuled. It was composed of many lobules,
separated from one another by fibrous septa, and each
enclosed in a capsule. Some of them grew into thin-
walled cysts containing little or no fluid, their walls
being in contact. The growths in these cysts were in
some cases lobulated finely on the surface. In a few, they
resembled in colour and smoothness a mucous nasal
polypus, but were in consistence a little firmer. The
majority of the lobules showed on section a finely
lobulated appearance, the outlines of the little lobules
being crenulated, and some having a distinct cavity in
the centre. No communication could be traced between the various cysts, nor could any be traced directly to the nipple. Some parts of the growth were
hard, firm, and fibrous. There were no blood cysts, nor
any of the translucent looking material so common in a
sarcoma. In some of the nodules (felt elastic during
life) there were many elongated slit-like spaces with
smooth walls, the surrounding material being acinous.

Histologically the tumour was composed of spindle
cells and gland elements (adeno-sarcoma).

Case 35. *Cystic (duct) sarcoma of right breast ;*

early *symptom hæmorrhage from nipple; amputation.*
(Reported by Mr. Phillips.)—Eliza A., aged 45, a governess, was admitted into Lydia ward on January **8th,**
1884, under Mr. Bryant's care.   Patient was delicate,
but never had any particular disease.   Breasts have
been always tender, but no history of injury.   Menstrual periods regular.

In January, 1883, one year before admission,
patient first noticed a pale brownish discharge **from**
the right nipple.   It lasted three months, and was
increased during her menstrual periods.   In March
she noticed a lump, the size of a small marble, on the
upper and inner part of her right breast.   This increased visibly during her menstrual periods.   About
two months ago it began **to grow,** and since then her
menses have been irregular.

On admission.   Patient is a florid-looking woman,
and apparently healthy.   The upper and inner aspect
of the right breast is occupied by a "firm but not
hard" lump, measuring $2\frac{1}{4} \times 1\frac{1}{2}$ inches, and with the
longest axis vertical.   The lump is painful.   There is
no fluctuation.   The skin is not very movable upon
the growth.   A hard cord can be felt running down
from the growth to the nipple.   During pressure
upon the tumour a few drops of serous fluid mixed
with blood came from the nipple.   Under the microscope this fluid was found to consist of blood
corpuscles forming rouleaux.   A small gland the size
of a nut exists in the axilla.

The breast was removed, and found after removal
to have been the subject of cystic degeneration of the
gland.   Into some of the cysts sarcomatous growth
of the spindle-celled variety existed.   Plate VIII.
Fig. 5 was taken from this case.

19th.   There was primary union all down the
wound except at its lower part.   The uppermost
stitches were taken out.

26th.   Drainage tube taken out.

February 1st.   There is perfect union of wound.

9th.   Discharged well.

The temperature ran a normal course throughout.

Case 36.   *Cystic adeno-sarcoma of breast; excision.*
—Miss C., aged 46, consulted me on July 10th, 1885, for
an affection of her left breast, which had been coming
on for about six months, and was accompanied with
some pain.   The whole gland seemed to be indurated
and enlarged, and to the hand felt coarse and
nodulated ; below the nipple a hard mass was felt.
The nipple was natural, and discharged a blood-stained
fluid, more particularly after manipulation.   The skin
over the breast was natural, and the lymphatic glands
were not enlarged.

On July 11th the breast was excised, and a good
recovery ensued.

To the eye the breast was clearly undergoing
cystic degeneration, the whole gland being full of
small cysts.   In one lobe these cysts had much increased, and one of them contained an intracystic
growth.   The cysts had evidently a connection with
the ducts, since during life there was a nipple discharge, and after the removal of the gland fluid could
be squeezed out of the nipple.

Mr. Symonds' report on the case is appended.

Report of case by Mr. Symonds.—The breast was
found to contain a large cyst in the tissue just beneath
the nipple, measuring one inch across.   Projecting
into one end of the cyst was a rounded solid growth,
with an uneven papillomatous-looking surface.   This
extends beyond the limits of the fluid-containing
portion of the cyst as a solid cylinder, round which
the cyst wall is closely fitted, so that in a transverse
section the appearance is much that of a large vein
filled with a growth.   The section has a whitish-grey
colour.   No other cyst of any magnitude was seen,

but there were many small ones. The rest of the
breast looked fairly healthy. The tissue was hard,
but no definite tumour was visible.

Microscopic examination.—The solid growth is
composed of irregular glandular acini, lined by
short columnar cells, with a deeper layer of **a** less
regular form. In many places the cells are heaped
up, and then are smaller. The stroma is composed
of fine fibrous tissue, and contains here and there
a cavernous arrangement of vascular tissue. It is
very delicate in parts, and forms but a small portion
of the whole; in other places, however, especially
where the growth is connected with the wall and outlying breast tissue, there is much solid material. This
contains much fibre tissue, spindle, and a few myeloid
cells, and closely resembles a sarcoma. This **solid**
growth does not form a separate tumour in the breast,
for at no great distance from the cyst wall the ordinary
mammary tissue is reached. Such a structure as that
described might be looked upon as the starting point
of the growth, but it forms so small a part of the
whole, and reaches so short a distance into the
breast tissue proper, that the view seems not so
reasonable as the one stated below. The cyst wall
is lined with the same short columnar cells.

There are other cysts visible under the microscope
varying from a size slightly larger than a normal acinus
to $\frac{1}{8}$th inch across. The smallest are evidently dilated
acini and ducts. These are lined with rounded or
columnar cells, according to their source of origin. The
larger ones, of which two or three were found, contain, besides the granules, large round or oval cells
with nuclei, and these cysts are lined by peculiar large
cells. They exhibit an intracellular and intranuclear
network, have a more or less columnar shape altered
by pressure, and large prominent nuclei. They are set
directly upon a wall of close fibrous tissue containing

elongated nuclei. In places these cells **are** heaped up
and proliferating. They are undoubtedly the progenitors of the large granular cells found in the
cysts. Sometimes cells of this character are found
filling spaces in the glandular formations, and showing in great contrast to the dark, short columnar
cells, for they are themselves pale, except for the
nucleus and granular appearance. In these smaller
cysts there are also found glandular growths, connected
with the wall. Where attached to the wall there is
no solid growth in the neighbouring breast tissue.

So far as the description has been carried, it will
appear that there is no solid growth except that found
in the cysts.

Examining other parts **of the** breast, it **is found**
that everywhere some · change is going on, that there
is a widespread activity manifesting itself throughout
the entire organ, culminating in the formation of
cysts and glandular growths. Again, it must be
stated that nowhere is any highly nucleated **tissue to**
be found except on the margin of the large cyst
above referred to. The fibre tissue is dense, and
contains few nuclei, and everywhere fat is found,
so that in all the illustrations adipose tissue, where
not sketched, will be considered to be present close
to the acini.

The presence of the fat seems **a** good evidence of
the fact that we are dealing with a diffused change
in the breast, and not with a widely disseminated
tumour, a point insisted upon by Cornil and Ranvier.

The first departure from the normal structure is
seen in the acini. The cells are increased in number
and obscure the lumen, and the wall shows as a
thickened fibrous investment, with very few nuclei.

Next several acini fuse together by obliteration of
the intervening septa, a process seen in all parts of
the breast in varying degrees. The cells are smaller

and more irregular than the normal ones. In some places, but this is exceptional, fibrous tissue is produced at the same time, and we have the appearances of an alveolated stroma of fibre tissue, the spaces of which are filled with epithelial cells. In the centre of some spaces is seen a lumen, while in others the cells have accumulated to such an extent as to obliterate all trace of it.

Such are the various morbid appearances seen in this breast. The sequence of events it is not so easy to arrange, for though there are some small adeno-matous growths in cysts, I cannot follow the develop-ment of such out of the earlier changes detailed above.

It is important to compare this case with the sketches of the breast from Mrs. T. (case 1), where a pure cystic change is seen, and with that of Mrs. T. (case 20), where a pure adenomatous growth hangs by a pedicle to the wall of a cyst. Whatever view be taken of the sequence of events in this breast, the following conclusions seem justified :

1. That the change is a general one.

2. That there is no new solid formation outside of cysts.

3. That the departure from the normal begins in the epithelium of the ducts and acini, and proceeds either towards cyst formation or fusion of many acini, to one or to both of which changes new formation of gland tissue succeeds.

It may, I think, also be surmised, when we con-sider the appearances presented by the large cyst in case 1, and the connection between the intracystic growth and the cyst wall in this case, that the glandular formation arises by outgrowths from the cyst wall, and is a pure intracystic growth, and not a protrusion inwards of a tissue formed outside the acini and ducts. As in cases 17 and 20, the growth may be viewed as an adeno-sarcomatous papilloma.

T—25

Case 37. *Cystic sarcoma of right breast; ulceration of skin over it ; excision ; recovery.* (Reported by Mr. H. Dismorr.) — Dora G., an anæmic-looking woman, aged 53, was admitted into Lydia ward on December 16th, 1873, under the care of Mr. Bryant. She had noticed for many years in her right breast a lump, at first not larger than a nut. Her sister, while in bed, had put her elbow on the patient's breast whilst raising herself up. This gave her a great deal of pain, and the small lump then appeared ; it gradually increased in size, but did not trouble her much till about six weeks before admission, when she caught cold, and it enlarged rapidly ; the skin broke in two places over two prominent lumps ; the bridge between them gave way and united the separated sores ; these increased in size and discharged very much. On admission, the breast was the seat of a tumour the size of a cocoanut, which was somewhat nodulated on the surface ; the skin on the inner side of the nipple was red from capillary congestion ; on the outside there was an ulcerated surface five inches by four, which was in great part covered by a blackish slough. The edge of the skin, though adherent to the tumour, was not infiltrated by new growth. The glands in the axilla were enlarged, but there was no enlargement of the abdominal viscera, and the other breast was healthy.

On December 23rd she was put under the influence of chloroform, and the tumour was removed by two semi-elliptical incisions, with the enlarged axillary glands. The two edges of the wound were brought together and supported by strapping. There were no adhesions of the tumour to the deeper parts. A section of the tumour showed a large single cyst with definite walls, its cavity being filled with sessile growths of a lobulated form, tough consistence, and

gelatinous look. The microscope showed it to consist of a loosely fibrillated connective tissue spindle-
celled sarcoma and a larger number of round fatty
nuclei scattered through it.

After the operation a morphia injection was given,
which greatly relieved the patient, but she was very
sick the day after, and perspired freely for some two
or three days.

29th. The wound is looking very healthy, and
covered with granulations. There has been a little
diarrhœa, accompanied by pain in the abdomen.
Temperature 98·6°, pulse 78.

January 17th, 1874. The patient has gone on
improving; there has been a good deal of perspira-
tion, at night particularly, after which she complained
of weakness, but her appetite has been good. Six
pieces of skin were transplanted to the wound, four
of which seem to have taken. She gets up for an
hour or two in the day. The wound decreased in
size, and on the 27th was the shape of a triangle, each
side being about two and a half inches long.

29th. Eight pieces of skin were transplanted,
six of which were thought to have taken five days
later from this time. She went on capitally, and left
the hospital on February 9th, the wound having lessened in size half an inch in ten days.

This patient was known to have been well three
years later.

Case 38. *Cystic carcinoma of breast; excision.*
—Miss S., about forty-five years of age, consulted me
on December 19th, 1884, on the advice of Mr.
Bisshopp, of Tunbridge Wells, for a tumour involving
her right breast of more than two years' growth.
She was first seen by Mr. Bisshopp in November,
1882, when the tumour had been discovered some
months. It was then hard, but painless; it grew
gradually. When I saw her the whole breast seemed

to be involved in the disease, and was hard and nodular. The nipple and skin over the breast were natural. I regarded the case as one of cancer, and advised its speedy removal. Sir James Paget saw the case and gave a like opinion. On January 9th, 1885, the operation was performed, and the wound healed at once by quick union.

The lady is now (two and a half years after the operation) well.

Report of tumour by Mr. Symonds.—The tumour was situated beneath the nipple in the substance of the breast, and measured about one and a half by two inches. The section when fresh showed a pinkish colour, with many yellow spots resembling altered secretion, plugging ducts. There were, besides, many larger spaces with smooth walls filled with a yellowish-grey or darker thick fluid. The tumour was not limited by any capsule, but blended directly with the breast, and in some places with the adipose tissue. When a thin slice was cut and the yellow contents of the spaces removed, the whole closely resembled cavernous tissue, so numerous were the spaces and so scanty the stroma. Examined microscopically in water, the yellow material was found to be composed of cells of all shapes and in all stages of degeneration, the prevailing shape being an irregular columnar. Numerous fatty granules, and plates and rounded or irregular masses of carbonate of lime also were present.

The microscopical appearances are shown in the drawing below (Fig. 11), made under a low magnifying power. The spaces are of various sizes and shapes, are all of a tubular character, and most of them have a distinct lumen, occupied when fresh by the yellow material above described, and which has nearly all fallen out in the preparation of the section. The epithelium lining these spaces is arranged mostly in a double or triple layer, having a distinct dark line where, uniting, they limit the lumen, an appearance resembling the striated border of Brücke. In other spaces the epithelium is more abundant, and forms irregular elevations projecting towards the lumen, which is in some

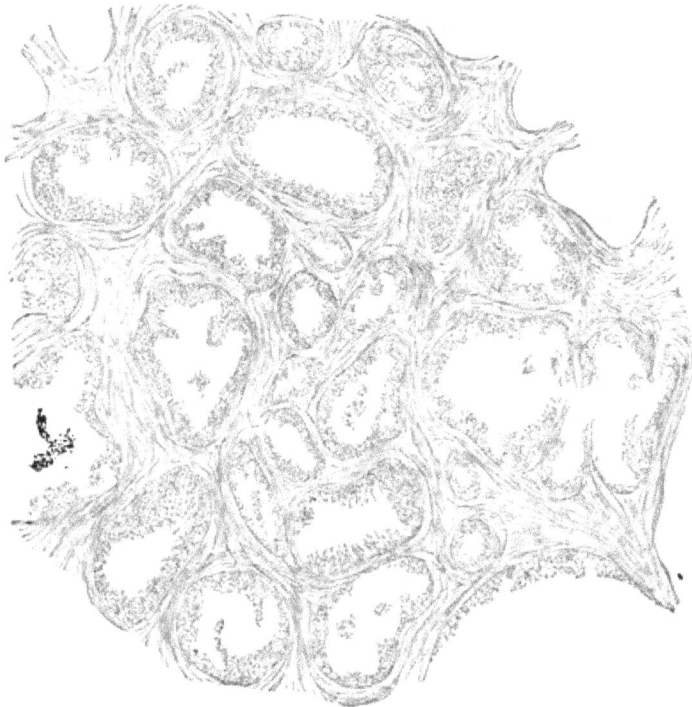

Fig. 11.—Tubular Carcinoma of Breast. Case 38.

instances entirely obliterated. Under a high power this last arrangement closely resembles an alveolus of an ordinary carcinoma of the breast. The stroma is fibrous, with very few nuclei or vessels, differing in these respects markedly from the stroma of new growths generally. The epithelial cells are large, angular, closely fit into one another, and have prominent round or oval nuclei. As they increase in number many become granular and some vacuolated.

The growth was not confined to the nipple, but was placed somewhat below this point. Moreover, it invaded the surrounding adipose tissue. On these grounds, and on account of the large size of the tumour, the large number of the spaces, and the absence of any large cysts, the formation must, I think, be considered as no mere duct dilatation, but a new growth, the prevailing character of which is its tubular form. This character is determined probably by the seat of origin, for when contrasted with another case, that of Mrs. T. (case 1), the absence in all the sections of ordinary breast tissue and fat confirms the view that this is a new formation.

Case 39. *Cystic cancer of breast.*—Miss B., aged 43, for ten years has had clear discharge from her left nipple, gluing it to her linen. At times the discharge has been blood-stained ; she has had occasional sharp pain in the breast. The catamenia regular.

Seven years ago the lump was the size of a nut ; now (April, 1882) it is the size of an orange.

' The lymphatic glands and the skin over the breast are healthy. Breast excised by Mr. Shipman, of Grantham, and found to be cancerous.

The growth had clearly grown from a cyst.

Case 40. *Cancer of breast following cystic disease coming on after prolonged suckling.*—Sarah C., aged 44, the mother of one child, which she suckled for three years up to six weeks ago, applied to me on April 25th, 1865, for a soft swelling the size of an egg in the outer lobe of her right breast. It was fluctuating, and had been coming two months.

Under tonics the swelling seemed to get smaller, but in June the skin over it became adherent, and in July pus escaped from the nipple.

On August 17th the cyst burst and discharged a yellow serous fluid. The case then did well.

In January, 1866, patient reappeared with the breast generally infiltrated and fixed to the chest; the skin over the breast was likewise infiltrated, and in one spot ulcerated.

At this time the diagnosis of an acute carcinomatous infiltration of the breast was easily made, and from the history of the case it seems probable that it originated within the cyst of which the report gives a history.

Had the cyst been excised when first treated, the cancerous disease might have been prevented.

Case 41. *Cystic carcinoma of breast following blood cyst; excision of breast; return and death in* **one** *year.*—Mrs. C., aged 58, nursed eight children without trouble. Lump in right breast for six **months,** having been preceded by a discharge from the nipple of blood for one and a half years. When I first saw her on May 5th, 1876, there was a cyst the size of a walnut on the outer side of her right nipple. By pressure all the sanguineous fluid could be squeezed out of it, and the wall of the tumour fell in, leaving a concavity. Pressure was applied to the tumour by cotton wool and strapping.

November 29th. No tumour can be felt. The hollow beneath nipple is still distinct. No discharge from nipple for three weeks.

July 28th, 1879. After the lapse of three years nipple retracting; some thickening in seat of cyst. No discharge from the nipple; skin slightly infiltrated over **seat** of old **cyst.** No axillary glands enlarged.

August 6th. Excised breast; good recovery.

June, 1880. Died with internal cancer.

The interest of this case rests on the fact that for at least four years before any symptom of local cancer appeared there was clear evidence of the presence of a duct cyst. That the cancer developed in the cyst there can be little question, since it was in the seat of the cyst and of the gland over it that the first signs of carcinoma appeared.

It was a source of regret to me when I saw this serious trouble, that excision of the breast had not been performed earlier. The results might then have been different.

Case 42. *Carcinoma involving a mamma the seat of cystic disease; excision;* **return** *five years later.*— Mrs. S., aged 42, the mother of one child, consulted me in February, 1876, for a tumour in her left breast which she had discovered four months, and which had steadily increased.

The tumour was situated in the axillary lobe of the gland; nipple natural; axillary glands free. **The** growth was clearly scirrhous.

Excision.—Axillary lobe infiltrated with carcinomatous material and simple cysts in other parts of the gland, one filled with inky fluid; no intracystic growths.

Good convalescence.

Remained well for nearly five years, that is till December, 1881, when signs of return of **cancer** appeared in the cicatrix.

Case 43. *Cystic carcinoma of breast following upon ordinary scirrhus.*—Ann C., aged 60, came under my care in July, 1864, with an infiltration of her right breast. Three months later the skin over the gland became adherent and infiltrated, and the axillary glands enlarged.

In April, 1865, a cyst appeared in the upper part of the tumour, and rapidly increased, so much so that in July it threatened to burst. This, however, did not take place. In September chest complications appeared which destroyed life.

Case 44. *Cystic carcinoma of breast ; excision ; cure.*—Ann C., aged 49, the mother of five children, all of whom she suckled.

For four and a half years she had discharge of blood-stained fluid from the nipple, which could be increased by pressure.

The gland had steadily enlarged during this period.

April 17th, 1865. Breast large and nodular. The skin was adherent to the breast, and the nipple was retracted ; the tumour was clearly cystic in its nature. The breast was subsequently excised and the patient recovered. The disease was, as suspected, cystic carcinoma, the gland being full of cysts, and the cysts of intracystic carcinomatous growths.

Case 45. *Cystic carcinoma of breast ; sloughing of growth ; hæmorrhage ; removal ; relieved.*—Harriet A., aged 70, was admitted into Lydia ward on January 28th, 1879, under Mr. Bryant. She has had four children, the youngest of which is now thirty-five years of age. There is no history of tumours, and patient has always been healthy.

The patient first noticed a small lump in the axillary portion of her left breast ten years ago, and it has continued to increase in size since. On Friday evening last it began to weep with blood.

On admission. The left breast is occupied by a hard infiltrating growth which implicates the skin and deeper parts, extending from near the left border of the sternum to the middle of the axilla, and from the third rib above to a point in a line with the ensiform cartilage. The breast is not distinguishable, and the site of the nipple is occupied by a depressed ulcerating induration marked with dilated vessels ; the ulceration is superficial, and covers an area an inch in diameter. External to this is a large, oval, fluctuating swelling, limited above by a deep furrow. The swelling is about three inches by two ; the skin over it is thin, red, and livid ; about the middle of it is a black spot, from which the hæmorrhage took place on the 24th inst. There is no enlargement of the axillary gland.

On the 28th and 29th more hæmorrhage.

January 30th. The cyst has collapsed ; the skin around the opening is sloughing, and it is discharging a fœtid grumous-looking material.

January 31st. *Operation.*—Two elliptical incisions were made from the median line of the body running obliquely downwards and outwards, about five inches in length, and three inches apart at the centre. The whole of the breast and carcinoma were removed as far down as the pectoralis major. The wound was washed with warm iodine water, two silver sutures were introduced at the upper angle of the wound, and a piece of terebene lint applied over it with a bandage to keep the dressings in their place.

February 2nd. The wound was dressed and the silver sutures removed.

March 10th. The patient was discharged to-day convalescent.

The walls of the cyst after excision on section showed small round cells. The hard mass was evidently scirrhus undergoing fatty degeneration. It was in one large cyst which had ruptured. There were other cysts filled with cancerous growths.

Case 46. *Cystic carcinoma of mammæ, with cystic degeneration of breast ; excision ; relieved.*— Mary Ann W., aged 50, was admitted into Astley Cooper ward on June 15th, 1878, with a hard swelling reaching from the nipple upwards two inches. The nipple was retracted and the skin was adherent to the growth beneath. The axillary lymphatic glands were not enlarged. Her previous history was good, as was that of her family.

It was only about six months before her admission

that she first noticed a little pain in the breast and a dimpling of the skin, beneath which was a small lump. The lump increased in size, and became very painful when touched. She then saw Mr. Bryant, who advised the removal of the breast.

June 16th. Chloroform was administered, and the breast removed. The mass was found to be cancerous, and cysts were found to be freely distributed throughout the gland tissue.

She left the hospital on Sept. 17th convalescent.

Case 47. *Recurrent carcinoma of mamma; excision; relieved; carcinoma of cyst.*—Jane R., aged 47, a married woman, was admitted into Lydia ward on 15th October, 1877. She was the mother of eight children, six of whom were living, the youngest being fifteen years of age. She had suckled all her children with her affected breast.

She had always enjoyed good health until about twenty months before admission, when she noticed for the first time a little lump in her right breast about the size of a walnut. Ten months later it increased rapidly.

On February 9th she was operated upon at another hospital, and the tumour was removed, leaving the breast and nipple, which were not infiltrated. She left the hospital on March 11th, with the wound **perfectly** healed.

On admission. The patient stated that about two **months** after her discharge from the hospital she noticed a slight swelling in the cicatrix of the old wound as well as pricking pains ; the growth rapidly increased in size until it became a circular tumour four inches in diameter, a little to the right of the sternum. It was hard to the touch, movable over the pectoral muscles, and infiltrated the skin for three inches. The skin was purple and shining, and had about its centre a small prominence.

October 23rd. Chloroform was administered, and the tumour was removed by two incisions, enclosing a space about five by six inches ; there was but little hæmorrhage. The growth had to be dissected off the pectoral muscles. The incision left a large cavity, into which a porringer without the handle might be embedded. This was left to granulate. One suture was, however, put in at the right-hand side of the wound.

The tumour, when cut into, exhibited a large cavity containing serous-looking fluid. The walls were composed of very soft cancerous growth made up of epithelial elements.

November 16th. There was a small growth, the size of a cherry, noticed to be growing from the bottom of the fossa near the lower and right side. This was thought to be a portion of the old growth left behind after the operation. The following day it was removed. There was very little bleeding and not much pain.

December 12th. She left the hospital convalescent.

Six months later this patient was well.

In this case the evils of a partial operation for carcinoma are well illustrated, for had the rule which is generally considered by surgeons as a binding one been observed, and the whole breast been removed when the primary tumour was taken away, the second operation would not probably have been called for.

Case 48. *Cystic sarcoma of one breast; removal; followed in nine years by cancer in opposite breast.*— Mrs. W., aged 60, the mother of one child, which she could not nurse as her nipples were retracted.

Twenty years ago she had a small lump in her right breast, which disappeared entirely in the course of two years.

February, 1872. Observed a lump in the same

spot, which has increased rapidly, so that now (April, 1873) it is the size of a cocoanut. Tumour clearly lobulated and cystic ; axillary glands free.

The breast was removed and was clearly cystic spindle-celled sarcoma ; a good recovery followed the operation.

June 14th, 1882. A tumour appeared in the left or opposite breast which infiltrated the gland, and gave rise to retraction of nipple and puckering of skin. It was an example of ordinary carcinoma.

Nothing was done, as the patient refused to be operated upon.

This case is worthy of record as an example of a patient having a carcinomatous tumour of one breast follow upon the removal of a cystic sarcomatous tumour in the other, although at ten years' interval. Such cases are rare.

I have, however, now under observation a lady, aged 68, who has had for at least twenty years a perfect specimen of scirrhous carcinoma of her left breast, which has remained unchanged for four and a half years, whilst in her right she has had a rapidly growing round-celled sarcomatous tumour removed with recurring growths on seventeen occasions, during these years, her general health being perfect. (*See* case 1, in chapter xx.)

Case 49. *Cystic carcinoma of breast.*—Mrs. R., aged 83, the mother of many children, all of whom she had suckled, consulted me on June 27th, 1877, for a cystic carcinoma of her left breast which had been coming on for one and a half years.

The tumour was of the size of an orange, and fluctuating ; skin over it infiltrated, but not the axillary glands ; nipple natural.

Maternal aunt had had cancer.

No operation advised on account of age.

Case 50. *Cyst in breast becoming the seat of*

carcinoma ; removal of tumour followed by death from secondary growth five years later.—The following case, which occurred in the practice of Mr. Birkett, is a good one to show how a cyst may exist for years in a breast gland before it becomes the seat of intracystic growth. The case is described as follows in the hospital catalogue (Preparation 2,309⁵): represents a

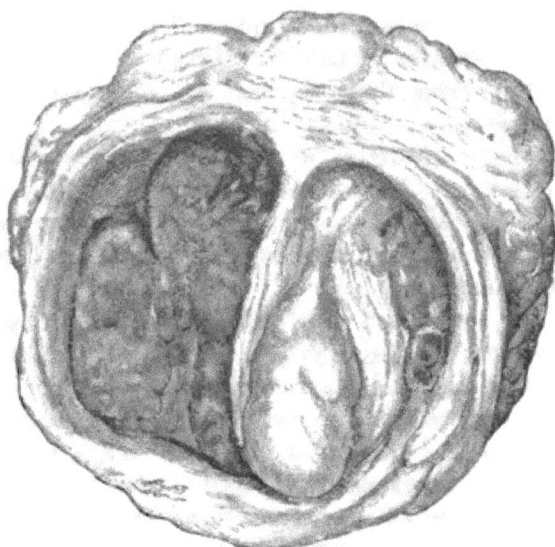

Fig. 12.—Cyst with pendulous Intracystic Carcinomatous Growth.
Case 50.

portion of the left mammary gland, consisting of cancer, which contained cysts. A large cyst (Fig. 12) is seen with a growth attached to its walls. The cyst was filled with bloody serum, and the growth consisted of the elements of carcinoma.

Case of Mrs. B., aged 40, a private patient of Mr. Birkett in 1853. Ten years before, she discovered a lump in her left breast, but suckled with it subsequently, and enjoyed good health. Five months before removal, on May 19th, 1853, the tumour became rapidly larger, and fluctuation could be detected in it. It measured more than three inches across, and occupied the axillary half of the breast.

The patient recovered from the operation, and was perfectly well for two years and nine months. A growth then formed in the axilla, which ulcerated, and cancer was developed in other organs, and for two years she suffered intensely. She died fourteen years after the discovery of the primary growth. On post-mortem, cancer was found in the left humerus, left femur, and left innominate bone, also on the lumbar vertebræ, and in the liver.

DIAGNOSIS OF CYSTIC TUMOURS.

In the examination of every tumour of the breast, and particularly of one in a woman over thirty-five, the surgeon as a matter of habit should allow the possibility of the case having a cystic origin to pass before his mind. By so doing he will not only prevent errors of practice which may be grievous, but at the same time he will guard against sources of serious mortification, for, I take it, there are few things more humiliating to a surgeon than an error in practice based upon a wrong diagnosis, when the error is not due to the inherent difficulties of the case, but to a want of care. When the tumour is in the breast of a woman over forty years of age this caution is more necessary, for cystic tumours of all kinds are more prone to occur in women over thirty-five and past the prime of life than in younger subjects. In 1861 I, however, removed a sero-cystic tumour from the breast of a girl aged twenty, which had been growing for three years, and at the time of operation measured twenty-two inches in circumference, and weighed after removal two and a half pounds. This preparation is preserved in the Guy's museum,* with another†

---

* **Prep. No.** 2295, drawing 480²⁰, model 40¹⁰.   † Prep. 2297²⁰.

in which a cystic duct sarcoma of one and a half years' growth was removed from a girl aged fifteen. When the tumour seems part of the breast gland itself, and appears as a single or possibly double growth in one or more lobes of the gland ; when the swelling is very hard and tense, and is not the seat of pain even when handled ; when it has existed for several years, and has given rise to no other symptoms than such as are referable to its mechanical presence, the probabilities of the tumour being cystic are very great. When the swelling is bossy or botryoidal, and fluctuation can be detected in some of its parts ; when with the swelling there is a discharge from the nipple, and this discharge can be increased by pressure or manipulation, the diagnosis of cystic disease is evident, and with the nipple discharge the diagnosis of a duct cyst is clear.

Whether the cysts be simple or proliferous is another question, and if the latter, the nature of the growth they contain is still a third. In a general sense it may be said that a single cyst is more probably simple than when the cysts are multiple and form a polycystic tumour ; that is, when a lobule of a gland is expanded by the presence of a single cyst, that cyst is more likely to be of a simple kind than when the cystic disease has involved more of the lobule. Simple polycystic disease of the breast is a rare affection, except as a result of general glandular degeneration. When a discharge from the nipple exists, and that discharge is bloody, the probabilities of a cyst containing some intracystic growth, however small, should be recognised, since it is from the vessels of the intracystic growth that the bleeding as a rule takes place. Simple cysts rarely bleed. When the bleeding is encouraged by or has followed manipulation, the probabilities of the cyst containing an intracystic growth are very great. When a cystic tumour exists in a

lobule of the gland as a fluctuating swelling in some parts and a more solid one in others, the diagnosis of solid with cystic growth is tolerably clear ; and when with this there is a free discharge of a bloody kind from the nipple, the diagnosis of cystic duct and glandular disease **is** highly probable. These lobular cystic tumours often present a **somewhat typical** shape, such as that of a pear, the narrow part **of which** is represented by the nipple (Plate VII. Fig. 1). Few of these tumours cause pain. Advice is sought as a rule either on account of the tumour or on account of the nipple discharge, not because of pain.

The **diagnosis of a cystic adeno-fibroma or adeno-sarcoma** is to be made out by attention to the same points as help the diagnosis of these growths when uncomplicated with cysts ; that is, a **cystic** adeno-fibromatous or sarcomatous growth simply troubles from its size or from its nipple discharge. By its growth it expands the integument over it, and spreads **out** such portions of the breast gland as may be superficial to it, but it in no other way involves its coverings (Plate VII. Fig. 2). The skin may be stretched even to a **rupture, but** even then the skin as tissue will be neither **invaded** nor infiltrated with disease. The cysts themselves may even burst and discharge their contents externally, but under such circumstances the integument that fringes the edges of the ruptured orifice will still **be seen** free, and can be readily distinguished even **from the** extruded or extruding intracystic fungating contents (Plate VII. Fig. 3). The nipple may be so flattened out as nearly to be indistinguishable, but it will only in very exceptional **cases be drawn in by** the disease, as seen so frequently in cancer ; and even should it be drawn in, inquiry will generally reveal the fact that such a condition was either congenital, **or that** some cicatrix of a former abscess had altered

U—25

its normal shape. Should the cyst be, however, more central and separate the lactiferous ducts, nipple retraction is to be expected.

The lymphatic glands (axillary and clavicular) will rarely be involved ; when they are they will be affected actively from lymphatic irritation rather than from lymphatic absorption, and under such circumstances the enlargement is more frequent when the **cysts** have burst through and irritated the skin than under other circumstances. **In** some forms of rapidly growing cystic sarcomatous as well as solid sarcomatous tumours, the lymphatic glands are, however, seriously affected.

The **diagnosis of cystic carcinoma** will turn upon the same points as that of the more common kind of cancer. Though the growth may have originated in a tube or cyst, it will not long be confined in its capsule, for, obedient to its nature, it will invade its cyst walls and creep into the tissue outside. The tumour **will** then spread into the neighbouring **parts by the** one marked way in which carcinomatous **tumours locally** spread, **viz.** by steady and progressive infiltration or **local** infection ; the carcinoma in its **progress** having first infiltrated the cyst in which **it** originated, then the tissues outside the cyst, later **on** those which surround these, **and** last of all **the integument** itself (Plate VII. Fig. 4). The infiltration **of the** integument **first** shows itself as a "dimpling" (Plate III.) of the surface, next as a "puckering" of the skin, thirdly as an "infiltration of the skin," and lastly as an ulceration. Should a cyst so enlarge as to encroach upon and infiltrate the skin, and later on rupture through it, **the** intracystic carcinomatous growth will project and **grow** through the opening, but the edges of the orifice through which the intracystic growth projects will be found everted (Plate IV. Fig. 2), and

involved in the disease, and not, as described in
the cases of cystic sarcoma, simply ruptured. The
infiltrating tendency of the disease is marked in
all its stages from its first to its last. The intracystic growth will never project and fungate in the
way of sarcomatous growths. Should the nipple be
affected by the disease, it may first be as a conduit
for the discharge through the milk ducts, but later
on, when the disease has left the cyst and invaded the
tissues outside its borders, it will be affected in other
ways. When the tumour by its growth has drawn
upon the suspensory ligaments of the gland and ducts
of the nipple, some retraction of the nipple will ensue,
and later on, should the nipple itself become materially
infiltrated, there may be actual projection. At times
there may be such infiltration and contraction of the
tissues around the base of the nipple as to produce
œdema of it from blood stasis, and in rarer cases even
sloughing.

The lymphatic glands, axillary or clavicular, **will**
become affected in the cystic form of carcinoma as
they are known to be in the non-cystic kind, and they
will be affected in the same way. In fact, the progress of a case of cystic carcinoma is similar **in** kind
to that of the more solid form of cancer. At the
beginning it may possibly be slower, but when the
disease has extended beyond the cyst wall or walls, I
am disposed to think it is often more rapid.

Prognosis. — A careful perusal **of the** fifty
cases quoted in this chapter (and I have published
all of which I have satisfactory records) will lead the
reader to form a very favourable opinion of the future
prospects of those who suffer from any cystic breast
trouble, and have been surgically treated. This re-
mark is applicable to every variety of cystic disease
with the exception of the carcinomatous. Where the
cyst is simple, a cure can safely be predicted by

surgical treatment, which may sometimes be of the simplest kind, such as merely puncturing the cyst and drawing off its contents. Sometimes it may be brought about by a free incision; in exceptional cases by excision. But by whatever means the cyst is abolished, a subsequent cure may be anticipated.

Whether it is wise to leave the disease alone is a question somewhat difficult to answer, but there is no doubt that in the cases recorded in which solid material was mixed with the cystic, the simple cystic element formed the beginning of the trouble; for in many of the cases a "lump" had existed in the breast for years (associated in some with a discharge from the nipple, which revealed its cystic nature) before any signs or symptoms of the presence of the more solid elements appeared. Under these circumstances it is not unfair to infer that if the simple cyst had not been allowed to remain, but had been treated, the more serious breast trouble which eventually attacked the different patients, and originated in the cyst, would not have manifested itself. If these remarks are applicable to the cases of simple cysts, they are clearly more so to those of a more complicated kind, such as the sero-cystic tumours. The prognosis in these cases after treatment is most satisfactory; indeed, a cure may with some confidence be predicted. In the four cases I have recorded the patients were known to have been well from six to eleven years subsequently, and, in fact, cured.

In the examples of cystic adenoma or sarcoma, the prognosis is no less good; indeed, if I am justified in taking my own cases as examples of all others, I might say the prognosis is very good, for of twenty-one cases recorded, sixteen of which were treated by excision, in no instances was there any known return of the local affection, and six of the subjects of the disease were known to have been alive after the operation for five, six, three, seven, eight, and twelve years respectively; whilst one died three years after having been relieved of her local trouble, from acute bronchitis. One patient with adeno-cystic sarcoma of one breast had a carcinoma of the opposite breast nine years later.

In cystic carcinoma a hopeful **view** may likewise be taken, for of my own nine cases which were treated by operation, one **was** known to have been alive two years, and another five years, after operation ; whilst a third remained well for nine years, when the opposite breast **became the** seat of an ordinary scirrhus. One other **case died** one year, and a second four years after **operation,** from internal cancer.

TREATMENT OF CYSTIC DISEASE OF THE BREAST.

In all cases in which the true cystic nature of the tumour is doubtful, an **exploratory** puncture with a small aspirating needle should be made, not only for diagnostic purposes, but with the reasonable prospect of retarding if not arresting the progress of the disease, and this hope may reasonably exist should the disease consist of the dilatation of a single duct, for I can recall not a few examples of what turned out to be simple serous cysts of the breast which had been tapped for diagnostic purposes in which, after the lapse of years, no return of the trouble has taken place. When a simple tapping has proved ineffectual or is inapplicable, an incision into the cyst, its free drainage, and healing by granulation, may bring about a cure, but these means should only be adopted when the cysts, on being laid open, have been found quite free from all signs of intracystic growths, and the cyst is single. When more than one cyst exists in a lobe this treatment is hardly applicable, and the safer practice is to dissect out the cysts with enough of the surrounding gland to make it reasonably probable that the disease has been wholly removed. When the cyst

contains an intracystic growth it must be freely excised, and should more than one such cyst be made
out, the whole gland had better be extirpated, for all
proliferating cysts are prone to grow, and the breast
glands of such as contain them are rarely only locally
implicated. I have, on many occasions, after the
removal of a cystic sarcomatous tumour, found such
portions of the gland as seemed to be healthy, the
seat of the same same disease only in miniature
(case 21), and this fact fairly suggests that the
disease, although palpable in one part and apparently
local, is really a general gland disease, and that under
such circumstances the whole gland had better be
excised.

In the more solid forms of cystic sarcoma and
cystic carcinoma, early excision is a rule of practice
which should never be deviated from.

CHAPTER XVII.

GALACTOCELES AND HYDATIDS.

Milk or lacteal cysts or galactoceles are comparatively rare cystic tumours of the breast. They
form when the gland is in a state of functional
activity, and usually during the first three months of
lactation. Exceptional cases are met with which appear at an earlier as well as at a later period. I shall
record one which appeared during the ninth month of
lactation. Atlee published in the American *Journal
of Med. Science* for April, 1874, an example which
developed sixteen months before childbirth, and increased rapidly during pregnancy; and Bouchacourt, as
recorded by Richelot, has reported an instance which

occurred in a woman, aged 51, twenty-four years after her last confinement.

As indicated by the name, the cysts always contain milk, cream, or a buttery material, which is clearly the product of milk ; and these variations in the character of the cystic contents are manifestly due to the changes which the fluid undergoes **after** it has been extravasated into the gland.

At times the tumour follows a blow or some inflammation, but as a rule it appears without any known cause during full lactation. The tumour is doubtless generally due to a rupture of the milk ducts, and consequent extravasation of milk into the connective tissue of the gland, although in some cases it may be caused by the obstruction and subsequent dilatation of a lacteal duct.

In one case the tumour will increase rapidly in **size,** and Birkett states that it may increase rapidly and distinctly during every time the infant sucks. In another it may appear suddenly, increase slowly, and then, as far as size is concerned, either remain stationary or even diminish, the diminution in size being occasioned by the absorption of the watery con-**stituents of the** milk, and its consequent concentration.

With respect to size, cases are on record in which several pints of milk have been evacuated. Scarpa is reported by Forget to have recorded an instance **in** which two pints of milk were removed from the tumour; and Birkett gives another from which, after excision, ten ounces of thick cream flowed away.

In the Guy's Hospital museum there is a preparation (2290[80]) of a large cyst which Mr. Birkett removed from the breast of **a married** lady, 27 years of age, which had been increasing for eight years. The cyst contained fluid composed of milk and serum, and in the walls of the cyst were several small adenomatous growths. It is probable that those growths were secondary to the formation of the galactocele.

At the same hospital (Preparation 2290[50]) there is also a breast removed from a middle-aged patient, for a tumour containing cheesy matter, occasioned by obstruction of a lactiferous tube. A thin translucent membranous expansion exhibits the boundary of the cyst, but all the contents have been removed. The tubes of the gland generally appear dilated, and pieces of glass rod are placed therein. There is no duct at present traceable to the cyst.

The *signs* and *symptoms* of this affection are more negative than positive. The one positive sign is the presence of a more or less globular swelling in a breast which is performing its physiological functions with full activity; one that may increase rapidly, or on the other hand, very slowly, and that gives rise to trouble mainly from its size. If it causes pain, it will probably do so from its weight. If it is attended with external evidence of venous congestion, as indicated by full veins, it will be so on account of its size. When the contents of the tumour are liquid, there will be fluctuation. When the fluid has been partially absorbed, and the contents are as it were condensed, the tumour will yield a doughy and peculiar feel. When the contents have become desiccated and a cheesy mass is left, it will present more the physical features of an adeno-fibroma or sarcoma. When it occurs during the period of suckling, and the child is put to the breast, pain is often produced. " The pain and distension (of the cyst) being increased by the draught of milk which enters the breast so soon as the child begins to suck " (Sir A. Cooper).

Should the lacteal cyst occupy the posterior part of the gland, and consequently be covered in front with a thickness of breast tissue, the globular swelling of the tumour will be obscured, and it may feel nodulated, the coarse, active gland tissue surrounding the cyst giving rise to this condition.

The nipple is not affected in this trouble, although at times it may discharge milk. When this symptom is present during the lactating period it is of no value, but when it exists after the breast has ceased this function, it is of great use; since it fairly shows that the cyst contents communicate directly with a duct, and thus suggests the diagnosis.

At times the tumours are multiple. I have seen two in the same breast. In this case the tumours were of the size of small eggs. They had appeared two years before I saw them, during lactation, and had for one year steadily diminished in size, even to a half, but since then they had been stationary, but felt doughy to the touch, and could be moulded to a degree like clay; that is, the side of one of the tumours **could be** made flat or convex by pressure. The second tumour could not be so altered, but it was placed more deeply in the centre of a lobe, and **had** the feel of a fibroma. I cut into both tumours, and evacuated **from** one what might have passed as excellent clotted cream, and from the other a cheesy material, which represented the cream still more condensed. This case was in a lady about 35 years of age, and the tumours had appeared during the lactation of her first child, **two** years previously.

A cystic tumour of this kind may therefore decrease in size, from absorption of its fluid contents, and as it decreases become more solid.

Treatment.—These cases, when met with in the months of early suckling, or when that function is in full activity, are not to be allowed to drift, for the milk will not be re-absorbed, and if **re-**absorption does not take place, the increase of **the** tumour may be expected as long as the functional activity of the breast continues. It is wise, therefore, should the swelling be large, to take the child **away** from the breast with the hope that with the disappearance of the secreting action of the gland, any increase of the tumour will be prevented, and its decrease encouraged ; when the cyst is small suckling may be continued.   Should this decrease in **size not ensue**, the cyst should be aspirated and its contents **drawn off.**   Should it then refill **and** the operation **prove** unsuccessful, an incision should be made into the cyst and its contents evacuated.   The incision should be a free one, for if a small one be made, it **will** soon close, and a re-accumulation of milk will take place, or a milk fistula will **be left.**   This milk fistula will last as long as the **breast is** active, and cease to discharge only when secretion has stopped.

The more chronic cases are to **be treated as** those of simple cysts, that is, by a free incision, the evacuation of the cyst contents and the subsequent closure of the cyst cavity by granulation.   **In the case** recorded, where solid contents existed, this method of **treatment** was successful.   In the following case, where the fluid was cream-like, a similar result ensued.

Galactocele followed by pregnancy and suckling of child from affected breast.—Susan T., 35 years of age, **who** was confined thirteen months previously, and suckled for ten months, came to me on October 5, 1865, with a **globular** tense swelling, the size **of** an orange, in her right breast, which had **been** coming four months, that is, from the sixth month of lactation. On November 20 the tumour was punctured for diagnostic purposes, and then freely opened, at least six ounces of cream exuding.   A good recovery ensued.

On September 25, 1866, this patient again came **to** me when she was pregnant nine months.   The breast **was** then apparently quite healthy.   She was confined on October 4, and suckled well.   No breast trouble **following.**

Galactocele of eleven years' standing, suppurating after fourth confinement; treated by free incision and cure.—Annie S., aged 31, the mother of four children, the youngest being nine weeks old, came under my care on April 1, 1874, with a tumour in the upper part of her *right* breast, which measured three and a half by two and a half inches in diameter. The skin over the tumour was adherent to it. The nipple was retracted.

Thirteen years before, when only 18 years of age, after a blow, she had pain in her right breast ; two years later she married ; when suckling her first child she noticed a small lump, the size of a walnut, on the upper part of her right breast. This never left, but after each of her confinements (four) increased in size. Between the periods it remained stationary. After her last confinement, nine weeks before admission, the lump increased rapidly, became painful, inflamed, and suppurated, some ounces of pus escaping.

She had never been able to suckle with the right breast, as the nipple was too flat.

A galactocele being diagnosed, I made a free incision into the tumour, and turned out a cupful of cream-like fluid, after which the cavity closed and recovery ensued.

Treatment.—The operation required is a free incision into the cyst in a line radiating from the nipple, followed by the thorough evacuation of the contents of the cyst and its perfect cleansing with iodine, or other antiseptic lotion. The cyst should subsequently be plugged with iodoform, carbolic, or other gauze, to promote healing by granulation. The cyst wall might, in very chronic cases, be well sponged with a twenty-grain to the ounce solution of chloride of zinc, or with the pure tincture of iodine before plugging. The wound should be left open to heal from below.

Hydatids in the Breast.

These cysts are found occasionally in the breast gland, as they are known to occur in any other part of the body. How they get there is a question which need not be discussed in these pages.

When such a cyst appears in the breast, it does so as a single, painless, tense elastic globular swelling of some portion of the organ unassociated with any morbid condition of the breast gland or nipple, or parts over the breast. It gives trouble merely mechanically by its size. After a period, and, it must be written, a very uncertain period, the hydatid will die, and under these circumstances the parts around will inflame and suppurate; the abscess eventually, if left to nature, opening and discharging; the dead wall of the hydatid in this way escaping. Indeed it is by the escape of the dead entozoon that the true nature of the case is sometimes for the first time revealed.

These cysts have been found in women of all ages, from twenty-three upwards. They grow slowly, at times very slowly; in two of the cases I record the tumour had existed for eight and eleven years respectively. In exceptional cases they attain the size of an orange rapidly.

They are without doubt difficult to diagnose from cystic disease of the breast, or some neoplasm, and more particularly when they are small. These tumours are, however, always single, and attain a larger size than most cysts; they are never associated with discharge from the nipple. Their true nature can, however, be readily made out by an exploratory puncture, and an examination of the fluid drawn off. The fluid of the hydatid being clear, or very slightly opalescent, alkaline, sp. gr. 1007 or 8, and *non*-albuminous; whereas the fluid of most, if not all, gland cysts is

PLATE VIII.

Fig. 1.

Fig. 2.

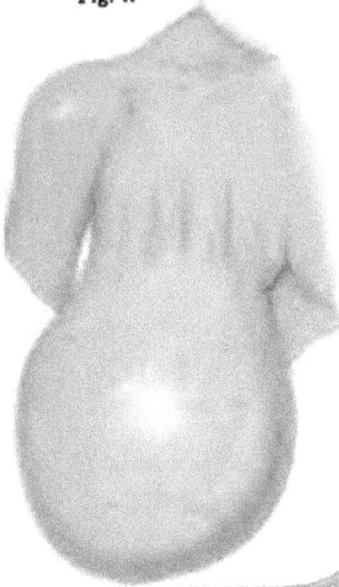

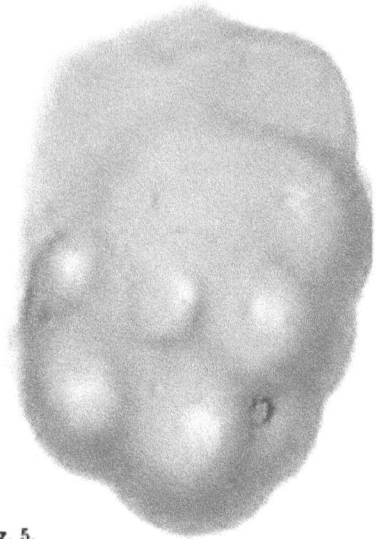

Fig. 5.

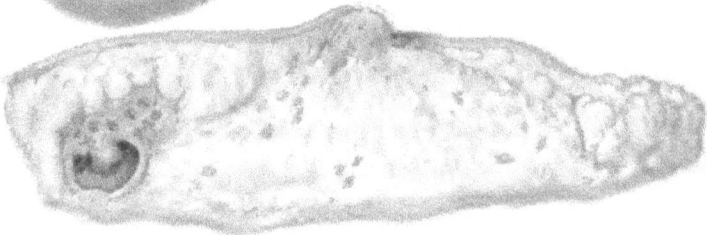

Fig. 3.

Fig. 4.

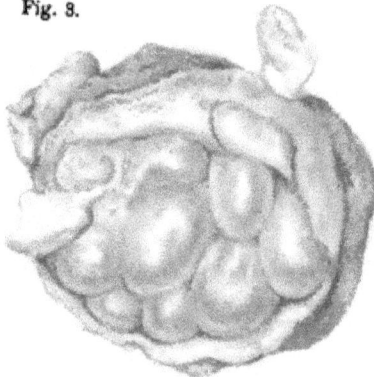

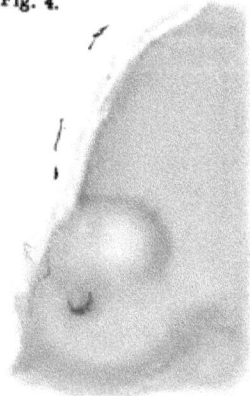

1. Lipoma.   2. Chondroma.   3 & 4. Hydatids.
5. Cystic Degeneration of Breast with sarcomatous Growths.

albuminous. When the characteristic **hooklets** are found in the fluid the diagnosis of any case is certain. Should the cyst hold a superficial position in the gland, its globular outline will be readily recognised ; in the case from which Fig. 4 on Plate VIII. was taken, this point is well illustrated. The case was one which **occurred** in the practice of **my** colleague, **Mr.** Symonds.

Hydatid of the breast.—Mary A. H., **aged 30, a** widow, came under my care at Guy's Hospital, in October, 1865, with a smooth, globular, fluctuating swelling, the size of a cocoanut, occupying the upper half of the left breast. She was a healthy woman, the mother of three children (the youngest being eight **years** old), all of whom she had suckled.

The tumour had been growing for five years, **and had** appeared as a small hard swelling, the size **of a** nut, above the nipple, and apparently deeply seated in the gland. It had grown steadily, although during **the** last twelve months its increase had been more rapid. It had never caused any **pain, and** had troubled her chiefly on account of its size.

The tumour and breast seemed to be one, and the gland moved freely over the deeper parts. The nipple was natural, and the skin over the breast was not implicated. There were no enlarged lymphatic glands.

The diagnosis, **in** this case, **was** not **easy** ; the swelling was evidently cystic, for all the symptoms indicating a cyst were well marked. The history **of the** case, **and** the almost total absence of pain and of all such symptoms as generally indicate suppuration, went far to prove **that** the swelling was not due to the presence of a chronic abscess. The nature of the swelling, and the apparently perfect healthiness of the uninvolved portion of the mammary gland, with the absence of nipple discharge, contra-indicated the

presence of ordinary cystic disease of the breast. The
origin of the swelling, as a small tense tumour, and
its gradual and painless enlargement, unaccompanied
by any definite symptoms of disease of the mammary
gland, were indications enough to excite a suspicion
of its hydatid nature, but the extreme **rarity** of such
an affection forbad any positive **diagnosis** being
made.

I therefore punctured the cyst with a trocar and
cannula, and drew off a few drops only of a thin watery
fluid containing flocculi of a delicate membrane, such
as was at once recognised as the lining membrane of a
hydatid cyst. On this diagnosis a free incision was
then made into the tumour, and out came a large
parent hydatid cyst, with many daughter cysts. The
whole measuring seventeen fluid **ounces.*** Under
the microscope, micrococci **were clearly** seen, and **the**
diagnosis established. The cavity from which the
hydatid escaped was left to granulate, and in three
weeks the woman was well.

I have given this case in full, with some remarks
I made with reference to diagnosis, when I originally
showed the hydatid tumour to the Pathological Society
on Nov. 7, 1865, for it contains within itself **an**
epitome of the history and progress of all similar cases,
and likewise illustrates the line of treatment which
should be adopted for diagnostic as **well as** curative
purposes.

In 1866 a woman was admitted under the care of
Mr. Hutchinson, into the London Hospital, for some
injury. It was then found that she had been twenty
years before in Guy's Hospital for some breast trouble,
and that a healthy cicatrix existed on the site **of her**
right breast. Mr. Birkett, to **whom** Mr. Hutchinson
then referred, found that this woman had been in 1846
a patient of Mr. Bransley Cooper, who had removed

* Prep. 2291⁵, Guy's Hosp. Museum.

her breast for a supposed tumour, which turned **out**
to be a hydatid.

The woman, when under Mr. Cooper, was a widow
of 51 years of age. She had had a swelling in her
breast for eleven years, and for eight it had been painless; for the last three years it had caused her incon-
venience, probably from its size. It was about three
inches in circumference, firm, and apparently solid.
On this account the breast was excised. After **its**
removal its hydatid nature was discovered. The
woman made a good recovery.

In the Guy's Hospital museum there is a preparation (2291) taken from a patient of Mr. Cooper Forster's,
which supports the points illustrated by my own case.
It was removed from a married, prolific, healthy woman,
aged 29, in the year 1856, who, six years previously,
had observed, when she was only 23 years of age, and
whilst suckling her first child, a small hard pea-like
swelling in the axillary lobe of her right breast. It
was painless, and had increased slowly. When admitted into Guy's the tumour measured four inches in
diameter, and was hard and elastic, but fluctuation
could not be detected in it. It moved freely with the
breast. Manipulation caused **pain.** It was explored
by an incision, and the parent cyst, containing
daughter cysts and limpid fluid, escaped. A good
recovery ensued. Fig. 3 on Plate VIII. illustrates
the case.

Occasionally, as already stated, these hydatids die,
and **when they** do, they excite irritation, inflammation, and suppuration. Suppuration is in a measure
curative, since it is the way nature adopts to get rid
of its unnatural lodger.

In Prep. 2291[10], Guy's Museum, a hydatid discharged by suppuration may be seen; it occurred in
the practice of Mr. Birkett, and I had the advantage
of seeing the case, the details of which are as follows:

Hydatid in the breast; suppuration; cure.—A married woman, aged 24, was admitted into Guy's in November, 1866, under the care of Mr. Birkett. She looked delicate, but stated that she had always enjoyed good health. She had, four years previously, given birth to dead twins, and had not since been pregnant; suckling, therefore, had never occurred. Eleven months before admission she accidentally discovered "a small lump" on the sternal half of her left breast, which slowly enlarged, but was unattended with pain, until about three or four months since. A month before admission the skin over the tumour became red, the tumour itself became larger, and more prominent; for the first time also it became painful.

When admitted the whole of the sternal half of the left breast formed a tumour, the skin covering which was red and painful when touched. The outward appearances were those of slow inflammation preceding suppuration, the skin here and there having a purple congested hue. After a few days, fluctuation in the tumour was clearly felt, and the skin began to ulcerate. Four openings altogether appeared, the largest being about half an inch in diameter. Thin, but fairly good pus escaped from the apertures for three or four days, when the nurse, as she removed the dressing, discovered the yellowish-white hydatid membrane, which had escaped from the largest opening in the breast, lying upon the dressing. After the escape of the hydatid the opening in the integument soon healed.

This case is an excellent one to illustrate a method that nature occasionally adopts to get rid of a hydatid cyst. In the present case the cyst was in the breast, but the method of cure would have been the same in other parts. Had the true nature of this tumour been understood, a more rapid cure would have been brought about by a free incision.

By way of summary it may be stated that hydatid tumour of the breast is generally discovered by accident, as a small, hard lump, embedded in one of the lobules of a healthy adult woman ; that it increases, as a rule, slowly and *painlessly*, and gives trouble solely by its size.   As it grows it becomes so identified with the breast tissue as to move with the gland, and to appear as part of its structure.   When it has attained the size of an egg, the sensation of fluctuation may be detected in it, and however much the integument may be raised by the tumour, it will always maintain its healthy appearance.

At an uncertain period of the cyst's life the tumour will become the seat of a subacute or chronic inflammation, which will end in suppuration, with the eventual discharge of the hydatid, and the cure of the case.   The period at which this result will ensue is, however, too uncertain to base a treatment upon. From the process employed by nature we can, however, learn our lesson as to treatment, and turn the cyst out of its bed through a free incision as soon as a diagnosis of the case has been made.   The cavity in which the hydatid rested will, on its removal, rapidly contract, and a recovery take place.

For diagnostic purposes in this as in the other cystic tumours to which attention has been directed, the use of the exploring needle cannot be too strongly advocated ; for by it, and it alone, can the true nature of the case be at once made known.

Sir A. Cooper originally described these cases, and illustrated them in his work on the breast (1829).

Haussmann of Berlin has recently given a good description of them.*

* *Medical Times and Gazette*, vol. ii. ; 1874.

CHAPTER XVIII.

A SUMMARY OF THE DIAGNOSIS OF TUMOURS OF THE BREAST.

To help the practitioner on the question of diagnosis of a mammary swelling, the following conclusions have been drawn up, which it is hoped may prove of assistance.

1. **Tumours that arise** during lactation **are** probably milk tumours, *i.e.* galactoceles or inflammatory swellings and abscesses.

2. Tumours that are found to be in, but not connected with the breast; that can readily be made out to be distinct from the gland, and moved without causing dragging upon the nipple, are presumably **of** the benign kind. If they are of slow growth, hard, inelastic, and lobulated, they are probably of **the adeno**-fibromatous variety; if of more rapid growth, smooth, somewhat elastic, and only slightly lobulated, adenosarcomatous; and if hard in parts, and soft in others, **clearly** fluctuating and bossy, they are probably cystic sarcomatous growths, or colloid.

3. A tumour that infiltrates a lobe or lobes **of the** breast, which cannot be separated from the gland **and** has no distinct boundary, is in its nature either inflammatory or cancerous; the lobe or **lobes** affected being in one case infiltrated with inflammatory products, in the other with epithelial elements.

4. When the affected breast has been physiologically active, or the seat of injury; when the swelling **is** ill defined and the mammary gland feels leathery, or painful and elastic, and when more than **one** of its lobes is separately involved, the **probabilities of** the affection having an inflammatory **origin** are very

great ; although when the infiltration has attacked
an inactive or obsolete breast, appears as a single
tumour, is hard, and nodular, the prospects of the
tumour being cancerous are reasonable ; and when, in
addition to these special local symptoms, there is either
"dimpling," "puckering," or infiltration of the skin over
the tumour, or the tumour with the **breast is fixed to the**
deeper structures, the diagnosis of cancer is confirmed.

5. Any globular, smooth tense tumour, situated
within and apparently forming part of a breast,
should be suspected to be of a cystic nature, and when
the tumour is associated with a discharge from the
nipple of a clear or blood-stained serum this suspicion
is much strengthened.

6. When more than one globular swelling is **pre-**
sent, or the breast feels coarse to the hand, the **gland**
is probably the seat of cystic degeneration **or of**
involution cysts. When the tumour is single and
there is no nipple discharge the tumour is either **a**
chronic abscess, a serous cyst, or a hydatid.

7. When the tumour is punctured for diagnostic
purposes, and **the** fluid withdrawn is brown, mucoid,
blood-stained, or blood, the cyst is probably of duct
origin ; and in proportion to the **amount** of blood in
the fluid is the diagnosis of an intracystic growth **to**
be made.

8. When the fluid is clear and albuminous, the cyst
is probably serous ; when watery and free from albumen **it** may with confidence be pronounced **to be**
hydatid. Under these circumstances the **character-**
istic hooklets will be found in the fluid.

9. A slowly growing tumour, which has shown no
signs of inflammation in its origin and progress, that
eventually becomes the seat of inflammation as indi-
cated by local redness, swelling. **heat,** and **pain, may**
be either a suppurating hydatid tumour, or a gum-
matous or tuberculous inflammation of the breast.

10. A solid or cystic tumour, however large, that simply distends the integument over it and has no tendency to infiltrate it, is clearly a solid or cystic adeno-fibromatous, or adeno-sarcomatous growth.

11. A solid or cystic tumour, however small, that gives rise either to dimpling, puckering, or infiltration of the skin over it, becomes fixed to the deeper tissues and is complicated with enlargement of the axillary or clavicular lymphatic glands, is certainly a cancer.

12. A *flattened* or *retracted nipple* associated with a tumour may be a symptom of small or great significance. If not congenital in its origin, or due to some antecedent inflammation, the *flattened* condition of the nipple may be brought about by a simple stretching of the gland, the result of continued growth of a simple neoplasm, whereas the *retraction* of the nipple may be produced either by the contraction of a scirrhous tumour infiltrating the lobe of the breast, and dragging upon its ducts, or by the presence of some adenoid, sarcomatous, or cystic tumour in the centre of the breast, and so separating its ducts as to bring about a drawing in and retraction of their terminations.

13. A *tumour that ulcerates* upon its surface and becomes excavated by the extension of the necrotic ulcerating process, is most probably cancerous, and when the edges of the ulcer are raised, indurated, and everted, the diagnosis is confirmed.

14. A tumour that presents a prominent fungating mass from some parts of its surface, and this mass projects from an orifice which has punched out and not infiltrated edges, is certainly sarcomatous, and probably cystic. A slow growing tumour which first stretches the skin and then ruptures it, and from the orifice of which thus made a colloidal or mucoid fluid escapes, is probably a colloid tumour.

15. A tumour which originated in the breast, that becomes complicated with a red or white, **brawny,** œdematous or tuberculated condition **of skin over** the growth is without doubt cancerous and **of the** worst type.

16. The absence of any enlargement of the *axillary,* or *clavicular lymphatic glands* with any breast tumour is an argument in favour of its benignancy, whereas the presence of such a complication suggests the reverse. Enlarged lymphatic glands may, however, be found associated with simple tumours when any local sources of irritation arise, and they may be absent for months, years. or altogether in certain examples of cancer, particularly of the atrophic variety in which the disease spreads slowly, and shows no sign of activity. In a case now under my observation of scirrhous cancer of fourteen years' standing, the lymphatic glands are uninvolved.

17. Discharge from the nipple **when free is** more than suggestive of a duct cyst ; where the discharge is serous, of simple serous disease ; where blood-stained or blood, of cystic disease complicated with intracystic growth, either of a simple or cancerous nature.

18. A slight sanguineous discharge from the nipple in the absence of nipple trouble, is suggestive of glandular cancerous disease, since simple non-cystic benign tumours never give rise to a discharge from the nipple unless associated with some degenerative **cystic** disease of the gland.

19. A slow growing, almost painless, nodular elastic tumour of **the** breast, over which the skin is thinly stretched before it becomes infiltrated and later on ruptured, and which discharges a tenacious mucoid fluid, more or less blood-stained, is certainly a colloid.

CHAPTER XIX.

ON MORBID CONDITIONS OF THE NIPPLE.

THE surgeon should in all affections of the breast be well alive to the fact that the nipple may be congenitally deficient, small, flattened out, or not prominent as it is normally. He should remember that a naturally well-formed nipple may have become deformed, or retracted from some inflammatory condition which occurred in early life, or followed a pregnancy which took place long before the complaint for which he may have been consulted appeared. He should know, moreover, that the nipple may have been destroyed by ulceration of a simple or malignant form, or that it may have sloughed off from some inflammatory trouble. He should, likewise, have clearly in his mind the fact that a nipple may have become flattened out or retracted from the growth of a simple benign tumour or cyst situated in the centre of the gland, and he should not jump to the too common conclusion when he finds a retracted nipple associated with a neoplasm, that the new growth is a cancer. A retracted nipple when associated with certain other morbid conditions is a valuable symptom of scirrhus, whereas by itself it has no significance.

There can be little doubt that as a positive indication of cancerous disease, the importance of a retracted nipple has been considerably overrated ; and that, although the symptom may be common in infiltrating cancer of the breast, such a disease may exist without it. It may be present, moreover, in simple non-cancerous affections. A retracted nipple may be regarded as an accidental symptom in the development of a tumour, and also as the product of mechanical causes; its presence being determined rather by the manner in which the gland is involved than by the nature of the disease. If **any tumour,** cystic or solid, simple or malignant, any abscess, chronic or acute, attack the centre of the mammary gland, a retracted nipple in all probability will be produced; for as a disease so placed necessarily causes material separation of the gland ducts, their extremities, terminating in the nipple, will be drawn upon, and, as a consequence, a retracted nipple **must** follow. We thus find this symptom of frequent occurrence in the early stage of an infiltrating cancer of the organ, the nipple being always drawn towards the side of the gland which may be involved; while **at** a later stage, **when** the infiltration is more complete, the nipple may again **project. In a** central chronic abscess of the breast, in **a** case of cyst **or of** adeno-fibroma or sarcoma, the retracted nipple **is** equally common, and in the true cystic adenocele it may be also present. In the ordinary adeno-fibroma **or** sarcoma, whether cystic or otherwise, it is rarely met **with,** for the reason that this disease is not of the breast gland **itself,** but only situated in its neighbourhood. In exceptional cases, however, **such an** association may co-exist. In one case in which I observed **it,** a blow or injury had preceded the **development** of the adenoid tumour, and it was open to a doubt whether the retracted nipple had not been brought about by some chronic inflammatory condition. **It** should always be remembered, moreover, that a contracted nipple may be a natural condition.

A discharge from **the** *nipple* should always attract attention. When the discharge is slight or of a bloody nature, it does not indicate any special affection, though it is well known that in *cancerous* affections a discharge from the nipple is not unfrequent, the fluid

having the appearance of blood-coloured serum, which
is never profuse, and rarely amounts to more than a
few drops. In the *true cystic* affection this symptom
is of considerable value, for in the majority of the cases
which have passed under my observation, as well as
in the recorded examples, this discharge from the nipple
was a prominent feature, the fluid being generally of
a mucoid nature, and more or less blood-stained;
and although at times it occurred spontaneously and
with relief to the patient, at others it could readily be
induced by some slight pressure upon the parts. In
the *ordinary solid forms of adeno-fibromata* or *sarcomata*, this symptom is seldom present. It exists
therefore as a symptom in the true cystic disease of
the breast structure, whether cancerous or sarcomatous; and is consequently, as a means of diagnosis, of
some value.

" The fluids," wrote Birkett, "which sometimes ooze
from the nipple at the commencement or during the
progress of a new growth, may be rendered subservient
to the formation of a correct diagnostication of the
nature of the disease in the part. Sanious, offensive
opaque discharges containing cells, identical with those
forming growths of cancer, may be regarded as in-
dicative that the induration which would probably
accompany the exudation of such fluid arises from infiltrating carcinoma; whilst a bright yellow, clear,
tenacious, serous fluid, drawing out into thread-like
processes, and the flow of which is perhaps increased
by compression on a circumscribed collection of fluid,
would guide the surgeon to an accurate opinion that
the tumour depended upon the presence of an adenoid
growth, or a simple cyst." *

**Inflammation and ulceration of the
nipples** is a serious trouble, since it is too often the

* Birkett: Holmes' "System of Surgery," vol. iii. p. 429.
Third edition.

precursor of mammary abscess. It is more likely to occur in first than in later pregnancies, but if women become feeble at the second, third, or fourth pregnancy, this affection may appear, although they may have escaped the trouble on former occasions. At times the affection shows itself as an inflammation of the nipple and its areola, with all the local phenomena of inflammation or swelling, redness, heat, and pain. In a certain number of cases the inflammation will subside under treatment, and the parts recover their normal condition.

In the majority of cases some local ulceration follows, and this may show itself either as a local "fissure," "chap," or "excoriation," or as a superficial or deep ulcer. In exceptional cases the nipple may be entirely destroyed. The ulcers are generally found between the rugæ or about the base of the nipple. In the worst cases the areola is likewise involved. The ulcerated surfaces are always painful and generally bleed. When the process of suckling is attempted the pain becomes agonising, and the hæmorrhage more abundant. That the pain should be severe under these circumstances can be readily understood. With these local symptoms there is usually much constitutional disturbance.

It has been asserted that these ulcers are commonly caused by some aphthous condition of the child's mouth; I believe they may be so in some cases, but in the majority they result from some unusual sensibility of the skin of the part, and at times from want of care. In first pregnancies the nipples should always be well looked to, and kept scrupulously clean, and if tender they should be bathed with some mild spirit lotion, or eau-de-Cologne and water, and well protected from friction of the dress by cotton wool or some glass or guttapercha shield. These shields help to make the nipples more prominent, and at the same time consequently prepare them for their function.

Where ulceration exists, soothing applications are the most valuable, such as Peruvian balsam, castor oil, or almond oil. At times glycerine **is** of use, whilst at others it causes pain.

When sore nipples occur at the time of suckling, shields should be worn. Great care should be observed **to** dry the nipples after the use of shields, and never to leave the shields in the child's mouth after the suck-**ling has** been completed. The application of the glycerine of tannic acid, Richardson's styptic colloid, tincture of catechu, a solution of nitrate of silver (gr. v to the ounce of water), or an ointment of extract of rhatany (gr. viij to ʒij of the oil of theobroma), are good applications. The summit of the nipple, whence the milk flows, should never **be touched** with **the** application. Whatever applications are used they should be washed off before the child suckles. In a few cases I have advised the use of a five per cent. solution of **cocain** before suckling, with great advantage. When cracks exist, it is a good plan for the mother to draw out the nipple by means of the old-fashioned feeding bottle before giving it to the infant, the mother's nipple being put into the central opening, and her mouth drawing the artificial one. Another ready method is the application to the nipple of the mouth of a wide-necked empty bottle that has been heated by hot water, the nipple, as the bottle cools, being pressed into the bottle and rendered prominent in a painless way. All breast pumps are to be condemned.

Eczema of the nipple and areola, or of the latter alone, is **a form** of dermatitis which must be recognised. It may appear as **an eczema in** other parts of the body, and get well by local treatment ; or persist, become chronic, and eventually take on the form of **what** Velpeau described as "eczema

rubrum," and pass into Paget's disease of the nipple,
the precursor of cancer, to which attention has been
already drawn.

This eczema may attack single or married women,
young or old ; the more chronic variety is generally
met with in the latter.   It is rarely confined to the
areola, but spreads to surrounding parts ; it often
attacks both breasts.

It is **best treated** by alkaline soothing application,
such as a lotion of the bicarbonate or borate of soda,
five grains to the ounce, with some extract of opium, to
get off the scabs, and then an ointment of zinc or lead,
separate or combined, or one of soda bicarbonate ten
grains to the ounce, or of boracic acid.   In the acute
stage, simple lead and opium lotion **is the** best.
Alkalies with **some** vegetable **bitter form** the best
internal medicine, but tonics and alteratives are often
required.   The diet **must** always **be** simple, and little
or no stimulants are, as a rule, required.

When the disease has spread to the end of the
nipple, and reached the orifices of the ducts of the
gland, it may creep into them, and it is possible that
in this **way** epithelioma may **arise ;** but this question
has already been discussed.

Follicular abscess of the areola is by **no**
means uncommon ; it may occur alone **or** in association with an eczematous or other form of dermatitis ;
it is to be treated as any other follicular inflammation,
and usually **soon** runs its course and gets well.

Ellen D., **aged 37,** the mother of four children,
the youngest **being** eight years old, came **to me** in
June, 1865, with suppuration of many of the follicles
of the areola of her right **breast.**   The disease had
existed for eleven weeks.   **By the use** of lead and
opium lotion and tonics, recovery **took** place.

Sebaceous tumours of the areola.—These
occasionally occur, and may give trouble if not

recognised. I have the notes of three such cases that came under my care when out-patient surgeon at Guy's, and they all occurred in young women; they may, however, be found in women of maturer age.

The cases are as follows:

On December 1, 1867, Mary G., aged 32, came under my care with a well-marked sebaceous tumour in the areola of her right breast. It had been coming for months. I turned it out of its bed as a whole by pressure through an incision, and the case did well.

Pedunculated sebaceous tumour of the areola.—On June 13, 1863, Susan D., aged 42, applied to me at Guy's Hospital, for a tumour hanging from the areola of the left nipple. It was pedunculated, and the size of a small nut. It had existed for three and a half years. When removed it turned out to be of a sebaceous kind.

Mary C., a married woman, aged 22, came to me on March 5, 1863, with a sebaceous tumour, the size of a nut, above and in contact with the nipple, which had been growing for five months. It had commenced during suckling. I removed the growth by turning it out of its bed. It was clearly sebaceous.

Charlotte S., aged 24, the mother of one child, consulted me on May 30, 1867, for an ulcerating sebaceous tumour on the margin of the areola of the right breast, which she had observed ten weeks. I enucleated the tumour, and the patient rapidly recovered.

These tumours may become the seat of cancer. I remember an example of epithelial cancer of the areola of a breast of a middle-aged woman, which followed a tumour which had existed for years, and broke down. I believed the case at the time to have been one of fungating sebaceous cyst; the interior of the cyst on its being emptied having become the seat

of epithelial cancer in the same way as is well known
"wens" upon the head or sebaceous cysts on other
parts of the body occasionally do.

Growths upon the nipple and areola.—
Warty, papillary, pedunculated, fibro-cellular, or other
growths may be found upon the nipple, which will be
recognised by their ordinary appearances.   I have

Fig. 13.—Pedunculated Papilloma of the Nipple.

seen many such, and give the particulars of two of
the cases, the notes of which I have preserved.

These growths should always be removed.   In
the Guy's museum there is a preparation (2,300⁶⁰) of a
pedunculated papillomatous growth, the size of a nut,
which was attached to the nipple of a woman, aged
48 (Fig. 13).   It had been growing for twenty-six
years, and during that time she had had ten children,
all of whom she had suckled.   Mr. Howse removed it.

*Pedunculated fibro-cellular growth on the apex of
the nipple the size of a nut.*—This was met with in
the person of Annie S., aged 27, a single woman, about
to be married, who came to me in May, 1864.   It

had been growing two years, and was clearly of the softer kind of fibro-cellular growths. I removed the growth with a .pair of scissors, and a good result followed.

Pedunculated growth the size of half a nipple growing from the lower border of the areola of the nipple.—Susan D., aged 42, came before me on June 18, 1863, with a pedunculated growth connected with the areola of her left breast, three-quarters of an inch long. It was firm to the touch, and apparently fibrous in texture. It had been growing for four years. She would not have it removed.

In the museum of the College of Surgeons (Prep. 4,819^A) there is a specimen of a pedunculated papillary tumour, nearly one inch and a half in length, which is smooth, firm, and lobulated. Microscopically the growth was composed of overgrown fibrous tissue ; but sections taken at its attachment showed an ingrowth of small well-defined epithelium, *not* like those of epithelioma. The growth was removed by Mr. J. Hutchinson from a woman aged 38.

In the same museum there is likewise a preparation presented by myself of an extremely atrophied breast only two and a half inches in diameter, of which the nipple is enlarged by a growth of dense fibrous tissue, both within and beneath it, and is pyramidal in shape. The other parts of the gland are of firm fibrous texture, but are less dense than the nipple. With the microscope, only fibrous tissue and a few compressed atrophied ducts, and no cancerous growth, could be found. It was removed from an elderly lady.

Chancre of the nipple is an affection which must not be left out of mind when the surgeon is consulted for an ulcer of the part. I have seen several examples of the kind, with the most typical chancres, infecting and non-infecting. A chancre upon this tissue presents much the same local appearance

as one upon the external part of the male prepuce.   **It**
is to be treated in the same way.

The subject is mentioned here in order that the
student and practitioner may be reminded of the
possibility of the nipple being the seat of primary
syphilis, the mere recollection of which renders the
accurate diagnosis of a case more probable.

CHAPTER XX.

ON THE PRESENCE OF MORE THAN ONE NEOPLASM IN THE SAME SUBJECT, AND ON THE SHRINKAGE OF TUMOURS.

IT is not common to find in the same subject more
than one kind of neoplasm, or to meet with cases of
cancer having apparently more than one centre. I
have, however, the notes of a few such cases, in which
the breast, with other parts, has been involved. I
quote them simply as curiosities, and as material for
future **study,** and not **as** texts **for** speculative
opinions.

The first case is **a very remarkable one:**

Case 1. *Rapidly growing sarcoma of right breast,*
associated *with atrophic scirrhus of the left* **breast** *of*
twenty years' *growth, and lipoma of the* **side;** *recurrence* **of sarcoma,** *and repeated operations,* ***in a***
lady, now 68 **years** *of age.*—Mrs. T., aged 64, a very
healthy looking woman, consulted me in 1882 for an
ovoid, smooth, semi-elastic tumour, the size of a cocoanut, which occupied the position of her right breast.
She **was** the mother of three children, all of whom she
nursed with the left **breast, but** not with the right,
as it was hard and **gave no milk.** The tumour had

been growing for about eight months, slowly at first, but rapidly for the last three months. The breast was lost in the tumour; and the skin and nipple, though healthy, were stretched; there was no enlargement of the axillary glands. The tumour was fleshy and elastic, with a smooth outline. It moved with the breast upon the parts beneath. I regarded it as a sarcoma. In the left, or opposite breast, was a typical atrophic carcinomatous tumour, which had existed for sixteen years. The breast was contracted up into a stony mass; the skin over it was puckered, but not infiltrated, and was adherent to the breast gland; the nipple had retracted deeply into the breast. The left axillary glands were not found to be enlarged. The carcinomatous tumour gave no trouble, nor was it the seat of pain.

Over the left hip there was a lipoma, the size of a fist, of twenty-five years' growth. This was disregarded by the patient.

On Jan. 9th, 1883, I removed the tumour with the right breast. The tumour was succulent, and on section showed a homogeneous structure; a glairy fluid exuded from its surface, which was of a pinkish-grey colour. Under the microscope the tumour was made up of spindle and round cells, with but little fibre tissue.

The case did well after the operation, but a return in the neighbourhood of the scar soon took place, for which another operation was demanded. This in its turn was followed by a speedy convalescence, and again by the return of the growth. Up to the present time, July, 1887, four and a half years after the first operation, sixteen operations have been performed, and the patient's general condition is excellent. Repair follows each operation in a most satisfactory way, and beyond the necessary trouble connected with an operation, the general health of the patient suffers in

no way.   Neither the carcinoma of the breast nor the
lipoma has made any progress.

The dates of the sixteen subsequent operations are
as follows : July 19th, Oct. 24th, 1883 ; April 2nd,
July 2nd, Oct. 8th, 1884 ; Jan. 16th, June 20th,
August 13th, Oct. 29th, 1885 ; Feb. 10th, April 28th,
July 17th, Oct. 9th, 1886 ; Feb. 9th, April 13th,
July 6th, 1887.   Sometimes one growth has been
removed, at others several.   Every growth has had
a delicate capsule, and each one has been readily
enucleated.   Some have been in the fatty tissue of
the part, others have been in muscle ; some have
been in the glands.

The specimen removed on April 2nd, 1884, was
examined by Dr. Goodhart, who reported the " growth
is a mixed sarcoma, chiefly a spindle-cell sarcoma, but
there are some myxoma-like cells in several places."

Mr. Symonds examined the tumour removed on
July 2nd, 1884, and reported as follows :

" The tumour was oval, measuring $1\frac{1}{2}$ in. by $\frac{3}{4}$ in.,
perfectly smooth in outline, and covered by a thin
capsule.   No breast or other tissue was attached to it.
It was soft and elastic, of a grey colour, and showed
an almost uniformly homogeneous section.   It could
easily be broken down by pressure, but at two points
was somewhat more resistant.   The surface when
scraped yielded a thin gelatinous fluid, containing
cells of various shapes, the majority were oval, with
several nuclei, and many were spindle-shaped.   The
nucleus of the latter closely resembled the former cell.
There were also a few round cells.

Microscopically the harder parts showed inter-
lacing tracts of spindle-cells, cut in various places,
giving the alveolated appearance common to such
tumours.   The vessels were numerous, the arteries
having well-defined walls, while the veins appeared as
spaces in the tissue, and being in places collected in

w—25

groups, gave a cavernous aspect to this part of the section.   In the softer parts the vessels were more numerous but smaller, the cells also smaller and more of them round, the whole having the appearance of granulation tissue.   The typical cell was a spindle, with a large oval granular nucleus.   The spaces referred to above as venous were of various shapes, and suggested, from their arrangement, that they communicated.   Their walls were very thin, and showed elongated nuclei in a small amount of imperfect fibrous tissue, but, as stated above, appeared more as spaces in the growth.   Though this is the common structure of veins in new growths, the arrangement in groups or plexuses seemed to call for special mention, as the spaces might be mistaken for cysts.   Their resemblance to the singly placed apertures, and the absence of any epithelial lining, confirms the interpretation here given."

Mr. Eve, of the College of Surgeons, examined the tumour removed on Feb. 10, 1886, and reported that it consists entirely of round cells in mucous tissue. " Probably," he adds, " the original growth belonged to the class of tumours formerly known as recurrent fibroids.   The elements get more elemental with each recurrence as a rule."   In this matter Mr. Eve is doubtless correct.

It is a point of great interest to find in a woman, now 68 years of age, a most active growing and recurring sarcoma, side by side with an indolent atrophic carcinoma, and with her general health perfect.   It is likewise worthy of note that in this case, as in cases 2 and 24, recorded in the chapter on cystic sarcoma, the disease attacked glands that had failed to secrete milk.

Case 2.   *Carcinoma of the breast, one year; lipomata of groins thirty years.*—Caroline B., aged 57, the mother of one child, came to me on Feb. 21, 1867, with a scirrhous infiltration of her right breast and adherent skin over the tumour which was fixed to the parts beneath. The axillary lymphatic glands were likewise enlarged. The disease had been discovered one year. Two large pendulous lipomata, the size of cocoanuts, also existed in the groins, which had been growing for thirty years.

Case 3. *Carcinoma associated with an* **adeno-fibroma** *of the right breast; amputation of breast.*— Eliza C., a single woman, aged 49, came under my care in May, 1885, with a tumour in her right breast, which had been growing for about three years. When seen the lower and axillary lobes of the right breast were evidently infiltrated with some new material, and the swelling formed a tumour three inches in diameter. It was hard and nodular to the touch, and moved with the breast upon the parts beneath. The nipple was natural, skin dimpled; the axillary glands were not felt to be enlarged.

On June 6 the breast was removed, and on exploring the axilla some lymphatic glands were found enlarged and taken away. The patient did well, and left the hospital in one month.

On making a section of the tumour, one-third of the gland was the seat of a true scirrhus, and the gland tissue about it was the subject of cystic degeneration. Between the cancerous tumour and the nipple, and upon the surface of the gland, there was a second tumour, the size of a nut, which was encapsuled, and on section presented a coarse lobulated surface, and to the eye and microscope yielded the appearance of an adeno-fibroma. In the Museum of the College of Surgeons a specimen of a like kind has been placed by Mr. J. Hutchinson.

Case 4. *Carcinoma of the breast following* **epithelioma** *of the nose.*—Jane B., aged 63, a childless married woman, came under my care in 1861, with epithelial

cancer of one nostril. This I removed, and a good recovery took place. Five years later, that is on June 25,
1866, when she was 68 years of age, she came to me
with a hard infiltrating carcinoma of her right breast,
associated with a puckering of the skin over the breast,
which had been coming for eight months. The disease progressed very slowly ; one year later tubercles appeared in the skin over the breast, and the
axillary glands became enlarged ; six months later the
arm was œdematous, and she died towards the end of
1868.

   Case 5. *Carcinoma of the breast of twenty-five
years' growth, followed by epithelial cancer of the nose.*
—**Frances H.**, a widow, aged 72, with one child, came
to me on July 1, 1865, with a marked example of
atrophic carcinoma of her right breast of twenty years'
standing, which had been ulcerating for seven years.

   In February, 1871, that is, six years later, when
she was 78 years of age, she came to me again
with a marked epithelial cancer of one of her nostrils.
The breast trouble, which had then existed for twenty-
six years, had altered but little.

   Case 6. *Infiltrating carcinoma of breast, and
ulceration of the nipple ; preceded by discharge from
nipple associated with cancerous stricture of the œsophagus.*—Mrs. G., a thin **feeble** woman, aged 60,
came under my care on February 25, 1885, with a
central hard tumour infiltrating the whole of the left
breast, and a raw red ulcer occupying the position of
the areola and nipple, and enlarged axillary and lymphatic glands. The nipple had entirely disappeared by
ulceration. She was the mother of two children, the
youngest being 25, both of whom she suckled without
trouble. For eight or nine years previously she had
had a watery discharge from the nipple, and nine
months later the discharge was bloody. Two years
ago the nipple became sore, and this soreness spread ;

about this time a lump appeared in the centre of the breast.

As the soreness of the nipple spread, the nipple steadily disappeared, and the tumour of the breast increased.

For five years she had had steadily increasing difficulty in deglutition, and for some months had only been able to take milk with sopped bread. The smallest size bougie can be passed through a strictured œsophagus.

As no active treatment was possible, I did not see this patient again.

Case 7. *Lymph-adenoma of cervical glands, followed by acute brawny cancer of the breast, and subsidence of the swelling in the lymphatic glands.*— Sarah B., aged 49, a childless married woman, came under my care in 1875, with enormous enlargement of the lymphatic glands in both sides of her neck, and behind and beneath the jaw. The swellings were composed of tumours the size of eggs, which could be moved about beneath the skin. These glandular swellings had been gradually coming for fifteen years. The woman was feeble and leucæmic.

Four years later this woman came to me again with a brawny carcinoma of her left breast, which had commenced four years before as a small lump, which grew slowly, and steadily involved the gland as a whole. When seen, the breast was generally infiltrated and hard. The skin covering it was adherent to the breast and felt like brawn; it had, moreover, in different parts many distinct indurated cancerous tubercles. The lymphatic axillary glands were enlarged.

Oddly enough, however, from the first appearance of her breast trouble, the cervical lymph-adenoma steadily subsided, so that when coming under observation but very slight enlargement of the glands could be made out.

The woman **had a** red complexion and a watery eye, but her **urine** was natural. Her powers were feeble, and I learned later on that she sank from what was supposed to be internal cancer.

The association of these two **diseases is** very interesting, and the fact that the lymph-adenoma diminished as the carcinoma increased is remarkable.

On March 6, 1874, I was present when my colleague **Mr.** Durham removed from the buttock of a **woman,** aged **65,** an **epithelial** cancer the size of a **florin, from** whom Mr. Hilton had excised, twenty-**three years previously, the right** breast for cancer, and **the scar was then a good one.**

On the same day **Dr. S.** K. Fowle, of New York, who was visiting London, told me that in 1858 he saw the late Dr. Gross remove from the cicatrix of an operation in the breast of an old woman a carcinomatous tumour, the size of a walnut, of one year's growth; **Sir** A. Cooper having excised the woman's breast thirty years before for cancer.

I would also here draw attention **to the** case of colloid tumour of the breast, published at page 201, in **which the** removal of a colloid growth in the breast was, **eight** years later, followed by a typical scirrhous carcinoma of the opposite breast.

Case **8.** *Carcinoma of the left breast ; excision ;* **return** *in the scar nine years later ; four years after re-***moval** *of breast, the appearance of a melanotic sarcoma* **in** *a mole situated in the left axilla ; excision and no* **return ;** *uterine cancer* **eight** *years after the removal of the breast; and carcinoma of the left femur, fracture of bone, and death nine years after removal of the breast.* —Mrs. K., aged 50, the mother of seven children, all of whom she had suckled without difficulty, the youngest being sixteen years of age, consulted me in May, 1876, with **an** infiltrating carcinoma of her left breast, which **had been slowly** growing for two years. The nipple was retracted, and the skin about the **nipple** infiltrated.   No axillary glands were enlarged.

On May 3, 1876, the breast was removed, and a rapid recovery followed.   The axilla was not explored, as nothing could be felt through the thin integument.   Four years after the operation in her breast, a mole, which had existed in her left axilla, on the side from which the carcinoma had been removed, became the seat of a melanotic sarcoma.   After six months' growth, when it had attained the size of a nut, I took this away and no return took place.

In 1884, eight years after the operation **on** her breast, the neck of the uterus became the **seat** of cancer; the disease appeared as a nodule **the size** of a hazel-nut, and from this there was at times free bleeding.

About this time the left arm became œdematous and painful.   In December of 1884 she complained of pain and stiffness in the upper part of her left thigh, which soon became swollen, and before many weeks had passed it was evident that the upper part of the femur was enlarged.   In June, 1885, some carcinomatous tubercles appeared in the scar of the breast operation, and when getting into bed she **felt some**thing give way in her thigh, and heard a crack.   Dr. Curgenven, her medical man, saw her, and discovered a fracture at the seat of the swelling, and there can **be no doubt** as to the swelling being a cancerous **tumour.**

From this **time** she steadily sank, and died in **August,** 1885, nine years after the removal of a carcinoma of the breast, and five years after the excision of a melanotic sarcoma of the axilla.

Case 9.   *Melanotic sarcoma originating in a mole situated in the right axilla, followed in four years by an infiltrating carcinoma of the right breast.*—Mary P., aged 40, a single healthy woman, came to me in April,

1872, with an ulcerating melanotic sarcomatous growth, the size of an orange, which had originated two years previously in a mole in the right axilla. No enlarged lymphatic glands were to be felt. On April 25th I excised the growth, and a good recovery followed. The patient remained well for nearly four years, when the breast of the same side became painful and enlarged, and six months later, when she came to me for advice, the breast was evidently the seat of an infiltrating carcinoma. The scar of the operation for melanotic sarcoma was sound, and no signs of recurrence were present. In July, 1876, I removed her breast, which on section was clearly of a carcinomatous nature, and a good recovery ensued. In 1884, eight years after the removal of her breast, and twelve years after the excision of the sarcoma, this patient was known to be quite well.

Shrinkage of tumours.—Adeno-sarcomatous and fibromatous tumours, in very exceptional occasions, diminish in size by time. I have certainly seen two examples of the kind and am disposed to think that the diminution in the size of the growth is to be explained by the absorption of such fluid as might have filled the growth or its capsule, rather than to any true shrinking of the neoplasm itself. Explain the matter how we will, the fact must, however, be recognised.

The cases are as follows :

Case 10. *Adenomata in the breast of a suckling woman, which diminished in size subsequently.*—Eliza F., aged 22, a married woman, pregnant four months, came to me on March 15th, 1868, with an adenomatous tumour, the size of half a cocoanut, in the axillary border of her *left* breast, and a second the size of an orange in the sternal border of the same gland. The two had been growing steadily for two years. During the progress of the pregnancy both these tumours grew rapidly, and became at last twice the size they were.

On Sept. 30th she was confined, and suckled with her right, or unaffected breast, for three months. The tumours in the left still grew. After weaning they, however, steadily diminished, so that **on Jan.** 3rd, 1870, that is, twelve months later, they had returned to the size described when seen in March, 1868.

It should be added that this patient's single sister, when 21 years of age, came to me in June, 1870, with great enlargement of her left breast, from supposed hypertrophy. The gland was twice the size of the right breast.

Case 11. *Adenoma of the breast of eighteen years' growth diminished by time.*—Margaret K., aged 41, a single woman, came under my care on July 3rd, 1873, with a tumour in her left breast of eighteen years' standing, which was the size of an egg, soft, lobulated, and movable in and with the breast. It was quite painless. Eight years previously the tumour had been more than twice its present size. It had grown steadily for two years, when growth stopped and changed into shrinkage.

.

CHAPTER XXI.

NÆVI AND VASCULAR TUMOURS OF THE BREAST— LIPOMATA AND CHONDROMATA OF BREAST.

NÆVI are found in the skin covering the breast, in the same way as they are found in other parts of the body. In **rare cases** they may be found upon the nipple. I saw an example of this many years ago in a female child, who had a red, raised, florid nævus, the size of a shilling, situated upon the **breast**

and half the nipple.   I destroyed it by means of the
galvanic cautery, after two or three applications.

Conrad Langenbeck has recorded cases of **nævi**
which have extended from the **skin to** the breast
beneath.   Such cases in infants must be difficult to
diagnose, since the breast as a gland is **but rudi-**
mentary, and even when the nævus involves **the**
deeper tissues, it may be a question as to how **far the**
breast may be involved.

That the breast may be the seat of true nævoid
disease there is no doubt, although such cases must be
rare.   I cannot find any examples recorded, beyond
the one I have briefly alluded to in my work on the
practice of surgery.   In that case the breast, with the
skin over it, was one vascular half globe.   It **was**
spongy to the hand, and could be emptied by pressure
and easily refilled.   The nipple was apparently **re-**
tracted, but this condition was brought about by the
elevation of the parts about it.   The notes of the case
as taken at that time are as follows :

Nævus involving the whole of the breast.—Mary R.,
fifteen weeks old, was brought to me at Guy's Hospital
on Oct. 14th, 1869, with a nævus involving the whole of
the breast gland and skin **over it ;** the nipple **was re-**
tracted.   The tumour measured two inches in diameter,
and was **like half a globe ; it was** spongy to the feel
and prominent.   Large **veins** converged **toward its**
surface and through the **skin.**   At **the** margin of **the**
gland the purple **substance** of the nævoid structure
was visible.   The **whole swelling** could easily be
reduced by pressure, and on the removal of which the
swelling returned.   **No** treatment was adopted.

Lipomata of the breast.—That lipomatous
tumours may be found in the integument that covers
the breast, that is, in the paramammary tissue, the
experience of most surgeons will prove, although it
is not so clear that they are **met** with within the

gland structure or behind it. Gross says, " I am **not** aware of a single case of circumscribed lipoma occur- **ring** in the gland itself," and if he means by this a fatty tumour unconnected with the surrounding fatty tissue *he is probably right,* **and his experience most** certainly coincides with my **own.**

Sir **A.** Cooper in 1805 removed the breast of a woman which measured in circumference thirty-one inches, **with a** large lipomatous tumour, which **was** subsequently found to be situated *behind* the gland, and to have weighed 14 lbs. 10 oz.

He also removed from a woman who had a breast of great size, **by a** single incision, masses of fatty tissue from between the different portions of the gland which was itself healthy. "The lobes of fat which are interspersed between the different portions of the mammary gland, and which serve naturally to augment **the** size of the bosom, having become enlarged and formed a swelling, which, prior to the incision being made, seemed to involve the whole of the breast ; but when the operation was performed, the different lobes of adeps which formed the tumour could be drawn away from the gland itself."

In the Guy's Hospital Museum (Prep. 2300[50]) there is preserved "a large adipose tumour, which measured 23 inches in circumference, and was removed after **death** from the breast of a married but sterile old woman, aged 87. When 30 years of age she observed the tumour in the upper part of the right breast, and this **she showed to** Sir A. Cooper on Jan. 9th, 1806. She remembered the date, because it was the day on which Lord **Nelson** was **buried.** **The** growth slowly enlarged, but not during the last few years. It was thought to have been an adenocele, but when a section was made it was found to consist of fat, with a piece of bone in the centre." Fig. 1 **on** Plate VII. was taken from this patient during life.

A lipomatous tumour may, however, be developed in the tissue covering the breast or elsewhere. The following is an example of the affection :

Lipoma the size of half an orange over the breast ; excision ; cure.—Annie B., aged 50, came under my care on Sept. 5th, 1883. She had had four children, the youngest being eleven years old. Five years ago she noticed a lump in the upper part of her left breast, which has steadily increased. When seen the lump was the size of half an orange, and measured three by two and a half inches. It was placed over the clavicular lobes of the left breast, and was movable over the gland. It was clearly lobulated and apparently fatty. On Sept. 7th I excised it and found it to be a pure lipoma. The case did well.

Chondromata.—These growths are rarely met with in the breast. **Sir A.** Cooper gives one case, with a drawing, in his work on the breast ; he removed it from a healthy woman, aged 32, and it was of fourteen years' growth. The pain in it was very severe, and the tumour was excessively hard. He removed the growth, and the larger portion of it had the appearance of that cartilage which supplies the place of bone in young subjects. The remaining portion was ossific. Fig. 2 on Plate VIII. illustrates the case.

Professors J. Müller, Rindfleisch, and Billroth, record other cases in which nodules of cartilage were found in the breast. In the Guy's Museum, Prep. 2316⁵⁰, there is a tumour which was taken from the breast of a middle-aged woman. It was examined by Dr. Wilks, who pronounced it to be fibro-cartilaginous.

In vol. xxxiii. of the Pathological Society's Transactions of London (p. 306), Mr. Bowlby reports a case of chondro-sarcoma of the breast, of one year's growth, taken from a woman, aged 42. The breast with the tumour, which was encapsuled, was removed, but a return took place which proved fatal within six months of the operation.

Cruveilhier has also described a case in vol. iii. of his "Anatomie-Pathologique," but Mr. Bowlby throws some doubt about its true nature. Upon the whole, although it must be accepted as a fact that the disease may occur in the breast, it is certainly very rare. It may occur as an innocent growth or in association with a malignant one.

On May 18th, 1886, Mr. W. H. Battle showed the members of the London Pathological Society a specimen of osteo-chondro-sarcoma of the breast, which is very rare.

It was taken from the right breast of a widow, aged 73, who had had five children. The tumour commenced as a hard lump, to the inner side of the nipple, six years before. The growth was painless, and the patient's general health unaffected. It grew slowly until, within a year of its removal, it had attained the size of a large orange; it consisted of two portions, an inner, very hard and rounded, about the size of a walnut, which the patient had noticed for a long time; and an outer, a more recent development, more elastic, of the size of a large egg. The nipple was much retracted. The skin was adherent, red, and tense at the inner part. One small freely movable gland was detected in the axilla. The growth consisted of a larger portion composed of a soft, friable, extremely vascular material, in which there had been numerous hæmorrhages; and of a smaller very hard portion which resembled bone, and could not be cut with a knife. Microscopic examination of the tumour showed it to consist, in the softer parts, of round and spindle cells; and in the harder, of cartilage which had in parts become ossified, the section showing well-marked Haversian canals.

CHAPTER XXII.

TUMOURS OF THE MALE BREAST.

THE male breast is liable to the same affections as the female, but in a far lesser degree ; **the** male gland at no period of its life being called upon to show any functional activity.

It may, in infancy, be as frequently the seat of **inflammation** as the female gland, and in a former **chapter it has been shown to** be liable about puberty **to** a marked growth, and to a spurious functional activity, which may lead up to inflammation. It may also, as a result of injury, be the seat of abscess.

Fibromata, sarcomata, and adenomata have been met with in the gland, and carcinomatous tumours are not rare. Thus Mr. W. R. Williams reports that out of about 280 examples of adenomata of the breast, consecutively noticed in London hospitals, one **was in a** male ; of 68 examples of sarcomata of the breast, two were in males ; and of 1433 cases of cancer of the breast, fourteen were in males ; or roughly speaking, that one case of benign tumour of the breast is met with in the male to 116 in the female ; and one case of carcinoma in the male to 102 in the female ; **car**cinoma in the female being at least a hundred times as common.

These averages are doubtless quite reliable, since they have been calculated from the combined experience of four large London hospitals, during ten, twelve, and seventeen years. They differ from smaller statistics published by Sir J. Paget, who believes that two cases of scirrhus of **the** breast occur in men to **98** in women ; by Gross, **who** has seen two examples in males to 100 in females ; **by** Billroth, who gives seven cases in males out of 252 cases; or taking all these cases **together, in** the proportion **of one to** forty-two.

Wagstaffe gives, in vol. xxvii. **of the** Pathological Society's Transactions (p. 246), **a table of** sixty-one **cases** of cancer of the breast in the male, collected from all sources, from which we learn that **in only six of** the whole number **the** disease appeared **under the age of 40; in twenty it** appeared between the ages of **40 and 60;** in seven between the ages of 60 and 70; **and in seven in** subjects over 70 years of age. The **disease** has evidently **a tendency** to occur **in** old rather than in middle-aged men.

The diagnoses **of** these **cases of neoplasm in the** male, **as** well as their **treatment, is to be determined by the same symptoms, signs, and conditions, as those of the** female.

INDEX.

x—25

PRINTED BY CASSELL & COMPANY, LIMITED, LA BELLE SAUVAGE, LONDON, E.C.

MANUALS

FOR

Students of Medicine

Published by CASSELL & COMPANY.

THIS Series has been projected to meet the demand of Medical Students and Practitioners for compact and authoritative Manuals embodying the most recent discoveries, and presenting them to the reader in a cheaper and more portable form than has till now been customary in Medical Works.

The Manuals contain all the information required for the Medical Examinations of the various Colleges, Halls, and Universities in the United Kingdom and the Colonies.

The Authors will be found to be either Examiners or the leading Teachers in well-known Medical Schools. This ensures the practical utility of the Series, while the introduction of the results of the latest scientific researches, British and Foreign, will recommend them also to Practitioners who desire to keep pace with the swift strides that are being made in Medicine and Surgery.

In the rapid advance in modern Medical knowledge new subjects have come to the front which have not as yet been systematically handled, nor the facts connected with them properly collected. The treatment of such subjects forms an important feature of this Series.

New and valuable Illustrations are freely introduced. The Manuals are printed in clear type, upon good paper. They are of a size convenient for the pocket, and bound in red cloth limp, with red edges. They contain from 300 to 540 pages, and are published at prices varying from 5s. to 7s. 6d.

Elements of Histology. By E. KLEIN, M.D., F.R.S., Joint-Lecturer on General Anatomy and Physiology in the Medical School of St. Bartholomew's Hospital, London. **6s.**

" A work which must of necessity command a universal success. It is just exactly what has long been a desideratum among students."—*Medical Press and Circular.*

Surgical Pathology. By A. J. PEPPER, M.S., M.B., F.R.C.S., Surgeon and Teacher of Practical Surgery at St. Mary's Hospital. **7s. 6d.**

" A student engaged in surgical work will find Mr. Pepper's 'Surgical Pathology' to be an invaluable guide, leading him on to that correct comprehension of the duties of a practical and scientific surgeon which is the groundwork of the highest type of British surgery."—*British Medical Journal.*

Manuals for Students of Medicine (*continued*).

Surgical Applied Anatomy. By FREDERICK TREVES,
F.R.C.S., Surgeon to, and Lecturer on Anatomy at, the London
Hospital. *New and Extended Edition.* 7s. 6d.

"The author of 'Surgical Applied Anatomy' is an able writer, and is also an authority
on purely anatomical questions. There are excellent paragraphs on the anatomy of
certain well-known surgical affections, such as hip-joint diseases, constituting a feature
quite original in a work of this class, yet in no way beyond its proper scope."—*London
Medical Record.*

Clinical Chemistry. By CHARLES H. RALFE, M.D.,
F.R.C.P., Physician at the London Hospital. 5s.

"The volume deals with a subject of great and increasing importance, which does
not generally receive so much attention from students as it deserves. The text is concise
and lucid, the chemical processes are stated in chemical formulæ, and wherever they
could aid the reader suitable illustrations have been introduced."—*The Lancet.*

Human Physiology. By HENRY POWER, M.B.,
F.R.C.S., Examiner in Physiology, Royal College of Surgeons of
England. 7s. 6d.

"The author has brought to the elucidation of his subject the knowledge gained by
many years of teaching and examining, and has communicated his thoughts in easy, clear,
and forcible language, so that the work is entirely brought within the compass of every
student. It supplies a want that has long been felt."—*The Lancet.*

Materia Medica and Therapeutics. By J.
MITCHELL BRUCE, M.D., F.R.C.P., Lecturer on Materia Medica at
Charing Cross Medical School, and Physician to the Hospital. Con-
taining an account of the action and uses of all the important new
Drugs admitted into the Pharmacopœia. 7s. 6d.

"We welcome its appearance with much pleasure, and feel sure that it will be
received on all sides with that favour which it richly deserves."—*British Medical
Journal.*

Physiological Physics. By J. McGREGOR-ROBERTSON,
M.A., M.B., Muirhead Demonstrator of Physiology, University of
Glasgow. 7s. 6d.

"Mr. McGregor-Robertson has done the student the greatest service in collecting
together in a handy volume descriptions of the experiments usually performed, and of
the apparatus concerned in performing them."—*The Lancet.*

Surgical Diagnosis: A Manual for the Wards.
By A. PEARCE GOULD, M.S., M.B., F.R.C.S., Assistant Surgeon to
Middlesex Hospital. 7s. 6d.

"We do not hesitate to say that Mr. Gould's work is unique in its excellence."—
The Lancet.

Comparative Anatomy and Physiology. By F.
JEFFREY BELL, M.A., Professor of Comparative Anatomy at King's
College. 7s. 6d.

"The book has evidently been prepared with very great care and accuracy, and is
well up to date. The woodcuts are abundant and good."—*Athenæum.*

A Manual of Surgery. Edited by FREDERICK TREVES,
F.R.C.S. With Contributions by leading Physicians and Surgeons.
Complete in Three Volumes, each containing about 600 pages fcap.
8vo, fully Illustrated. 7s. 6d. each.

Forensic Medicine. By A. J. PEPPER, M.S., M.B.,
F.R.C.S., Examiner in Forensic Medicine to the University of
London.

Hygiene and Public Health. By SHIRLEY F. MURPHY,
M.R.C.S., Lecturer on Hygiene and Public Health, St. Mary's
Hospital. 7s. 6d.

Other Volumes will follow in due course.

Cassell & Company, Limited, Ludgate Hill, London

CLINICAL MANUALS

FOR

Practitioners and Students of Medicine.

Complete Monographs on Special Subjects.

"A valuable series, which is likely to form, when completed, perhaps the most important Encyclopædia of Medicine and Surgery in the English language."—*British Medical Journal.*

THE object of this Series is to present to the Practitioner and Student of Medicine original, concise, and complete monographs on all the principal subjects of Medicine and Surgery, both general and special.

It is hoped that the Series will enable the Practitioner to keep abreast with the rapid advances at present being made in medical knowledge, and that it will supplement for the Student the comparatively scanty information on special subjects contained in the general text-books.

LIST OF CLINICAL MANUALS.

Ophthalmic Surgery. By R. BRUDENELL CARTER, F.R.C.S., Ophthalmic Surgeon to, and Lecturer on Ophthalmic Surgery at, St. George's Hospital; and W. ADAMS FROST, F.R.C.S., Assistant Ophthalmic Surgeon to, and Joint-Lecturer on Ophthalmic Surgery at, St. George's Hospital. With Chromo Frontispiece. **9s.**

Diseases of the Breast. By THOMAS BRYANT, F.R.C.S., Surgeon to, and Lecturer on Surgery at, Guy's Hospital. With 8 chromo plates. **9s.**

Diseases of Joints. By HOWARD MARSH, F.R.C.S., Senior Assistant Surgeon to, and Lecturer on Anatomy at, St. Bartholomew's Hospital, and Surgeon to the Children's Hospital, Great Ormond Street. With Chromo Frontispiece. **9s.**

"This volume is excellently planned. Mr. Marsh brings to bear upon it keen critical acumen."—*Liverpool Medico-Chirurgical Journal.*

Diseases of the Rectum and Anus. By CHARLES B. BALL, M.Ch. (Dublin), F.R.C.S.I., Surgeon and Clinical Teacher at Sir P. Dun's Hospital. With chromo plates. **9s.**

"As a full, clear, and trustworthy description of the diseases which it deals with, it is certainly second to none in the language. The author is evidently well read in the literature of the subject, and has nowhere failed to describe what is best up to date. A model of what such a work should be."—*Bristol Medico-Chirurgical Journal.*

List of Clinical Manuals (*continued*).

Syphilis. By JONATHAN HUTCHINSON, F.R.S., F.R.C.S., Consulting Surgeon to the London Hospital and to the Royal London Ophthalmic Hospital. With 8 chromo plates. **9s.**

" A valuable addition to the series of Clinical Manuals of its publishers, by an expert and accomplished writer, moderate in tone, judicious in spirit, and yet expressing the decided convictions of one whose experience entitles him to speak with authority. The student, no matter what may be his age, will find in this compact treatise a valuable presentation of a vastly important subject. We know of no better or more comprehensive treatise on syphilis."—*Medical News, Philadelphia.*

Fractures and Dislocations. By T. PICKERING PICK, F.R.C.S., Surgeon to, and Lecturer on Surgery at, St. George's Hospital. **8s. 6d.**

" We must express the pleasure with which we have perused the book, and our especial admiration for the lucidity of the author's style, and the simplicity of his directions for the application of apparatus; in the latter respect it is always difficult to combine clearness with brevity, but herein Mr. Pick has been most successful."—*Glasgow Medical Journal.*

Surgical Diseases of the Kidney. By HENRY MORRIS, M.B., F.R.C.S., Surgeon to, and Lecturer on Surgery at, Middlesex Hospital. With 6 chromo plates. **9s.**

" Mr. Morris writes clearly and forcibly, and handles his subject very thoroughly, so that the reader rises from the perusal of the work impressed with its importance. It would be difficult to find these subjects treated more carefully and thoroughly."—*British Medical Journal.*

Insanity and Allied Neuroses. By GEORGE H. SAVAGE, M.D., Medical Superintendent and Resident Physician to Bethlem Royal Hospital, and Lecturer on Mental Diseases at Guy's Hospital. **8s. 6d.**

" Dr. Savage's grouping of insanity is practical and convenient, and the observtaions on each group are acute, extensive, and well arranged."—*The Lancet.*

Intestinal Obstruction. By FREDERICK TREVES, F.R.C.S., Surgeon to, and Lecturer on Anatomy at, the London Hospital. **8s. 6d.**

" Throughout the work there is abundant evidence of patient labour, acute observation, and sound reasoning, and we believe Mr. Treves's book will do much to advance our knowledge of a very difficult subject."—*The Lancet.*

Diseases of the Tongue. By H. T. BUTLIN, F.R.C.S., Assistant Surgeon to St. Bartholomew's Hospital. With 8 chromo plates. **9s.**

" Mr. Butlin may be congratulated upon having written an excellent manual, scientific in tone, practical in aim, and elegant in literary form. The coloured plates rival, if not excel, some of the most careful specimens of art to be found in the pages of European medical publications."—*British Medical Journal.*

Surgical Diseases of Children. By EDMUND OWEN, M.B., F.R.C.S., Surgeon to the Children's Hospital, Great Ormond Street, and Surgeon to, and Lecturer on Anatomy at, St. Mary's Hospital. With 4 chromo plates. **9s.**

" Mr. Owen's volume will rank as an invaluable *résumé* of the subject on which it treats, and should readily take its place as a reliable and compact guide to the surgery of children."—*Medical Press and Circular.*

The Pulse. By W. H. BROADBENT, M.D., F.R.C.P., Physician to, and Lecturer on Medicine at, St. Mary's Hospital.

Other Volumes will follow in due course.

Cassell & Company, Limited, Ludgate Hill, London.

5

THE YEAR-BOOK OF TREATMENT.

A Critical Review for Practitioners of Medicine.

THE object of this book is to present to the Practitioner not only a complete account of all the more important advances made in the Treatment of Disease, but to furnish also a Review of the same by a competent authority.

Each department of practice is fully and concisely treated, and such allusions to recent pathological and clinical work as bear directly upon Treatment enter into the consideration of each subject.

The medical literature of all countries is placed under contribution, and the Work deals with all matters relating to Treatment that have been published during the year ending Sept. 30th. *A full reference is given to every article noticed.*

"This book is the combined work of twenty-three contributors, who have not only abstracted the best contributions to the practice of medicine and surgery during the twelve months, but have criticised them. The whole is compressed into 300 octavo pages, and the matter may be said to lie in a nutshell. The work appears to have been apportioned to the individual contributors with excellent judgment, and the result is *a book of extreme value* to all who in these busy times find it difficult to keep pace with the ever-advancing march of the science and art of medicine."—*Lancet.*

"This handbook contains, within the space of three hundred pages, a wonderfully complete summary—review of the methods of treatment, new and resuscitated, which have been advocated during the year with which it deals."—*British Medical Journal.*

7

Our Homes, and How to Make them Healthy.

With numerous Practical Illustrations. Edited by SHIRLEY FORSTER MURPHY, *late Medical Officer of Health to the Parish of St. Pancras; Hon. Secretary to the Epidemiological Society, and to the Society of Medical Officers of Health.* 960 pages. Royal 8vo, cloth . . . . . *15s.*
Roxburgh . . . . . . *18s.*

CONTENTS.

Health in the Home. By W. B. RICHARDSON, M.D., LL.D., F.R.S.

Architecture. By P. GORDON SMITH, F.R.I.B.A., and KEITH DOWNES YOUNG, A.R.I.B.A.

Internal Decoration. By ROBERT W. EDIS, F.S.A., and MALCOLM MORRIS, F.R.C.S. Ed.

Lighting. By R. BRUDENELL CARTER, F.R.C.S.

Warming and Ventilation. By DOUGLAS GALTON, C.B., D.C.L., F.R.S.

House Drainage. By WILLIAM EASSIE, C.E., F.L.S. F.G.S.

Defective Sanitary Appliances and Arrangements. By PROF. W. H. CORFIELD, M.A., M.D.

Water. By PROF. F. S. B. FRANÇOIS DE CHAUMONT, M.D., F.R.S.; ROGERS FIELD, B.A., M.I.C.E.; and J. WALLACE PEGGS, C.E.

Disposal of Refuse by Dry Methods. By THE EDITOR.

The Nursery. By WILLIAM SQUIRE, M.D., F.R.C.P.

House Cleaning. By PHYLLIS BROWNE.

Sickness in the House. By THE EDITOR.

Legal Responsibilities. By THOS. ECCLESTON GIBB.

&c. &c.

"A large amount of useful information concerning all the rights, duties, and privileges of a householder, as well as about the best means of rendering the home picturesque, comfortable, and, above all, wholesome."—*Times.*

Seventh and Cheap Edition. Price 1s. 6d. ; cloth, 2s.

A Handbook of Nursing

For the Home and for the Hospital. By CATHERINE J. WOOD, *Lady Superintendent of the Hospital for Sick Children, Great Ormond Street.*

www.ingramcontent.com/pod-product-compliance
Lightning Source LLC
Chambersburg PA
CBHW030858270326
41929CB00008B/468